NWP 3-05

NAVY WARFARE PUBLICATION

NAVAL SPECIAL WARFARE
NWP 3-05

EDITION MAY 2013

DEPARTMENT OF THE NAVY
OFFICE OF THE CHIEF OF NAVAL OPERATIONS

DISTRIBUTION RESTRICTION: DISTRIBUTION AUTHORIZED TO U.S. GOVERNMENT AGENCIES AND THEIR CONTRACTORS ONLY FOR OPERATIONAL USE TO PROTECT SENSITIVE TECHNICAL DATA OR INFORMATION FROM AUTOMATIC DISSEMINATION. THIS DETERMINATION WAS MADE APRIL 2013. OTHER REQUESTS SHALL BE REFERRED TO NAVY WARFARE DEVELOPMENT COMMAND, 1528 PIERSEY STREET BLDG O-27 NORFOLK VA 23511-2723

PRIMARY REVIEW AUTHORITY:
COMNAVSPECWARCOM

URGENT CHANGE/ERRATUM RECORD		
NUMBER	DATE	ENTERED BY

0411LP1136423

INTENTIONALLY BLANK

NWP 3-05

DEPARTMENT OF THE NAVY
NAVY WARFARE DEVELOPMENT COMMAND
1528 PIERSEY STREET BLDG O-27
NORFOLK VA 23511-2723

May 2013

LETTER OF PROMULGATION

1. NWP 3-05 (MAY 2013), Naval Special Warfare, is UNCLASSIFIED. Handle in accordance with the administrative procedures contained in NTTP 1-01 (APR 2005), The Navy Warfare Library.

2. NWP 3-05 (MAY 2013) is effective upon receipt and supersedes NWP 3-05 (SEP 2009), Naval Special Warfare. Destroy superseded material without report.

3. NWP 3-05 (MAY 2013) contains information on the nature, forces, organization, and employment of naval special warfare. It is intended for use by military planners and staff officers.

4. NWP 3-05 (MAY 2013) distribution is authorized to U.S. Government agencies and their contractors only for operational use to protect sensitive technical data or information from automatic dissemination. This determination was made April 2013. Other requests shall be referred to Navy Warfare Development Command, 1528 Piersey Street BLDG O-27 Norfolk, VA 23511-2723.

S. B. JERABEK

INTENTIONALLY BLANK

NWP 3-05

May 2013

PUBLICATION NOTICE

ROUTING

1. NWP 3-05 (MAY 2013), NAVAL SPECIAL WARFARE, is available in the Navy Warfare Library. It is effective upon receipt. NWP 3-05 supersedes NWP 3-05, Naval Special Warfare (SEP 2009).

2. Summary. NWP 3-05 contains information on the nature, forces, organization, and employment of naval special warfare. This publication is intended for use by military planners and staff officers.

Navy Warfare Library Custodian

Navy Warfare Library publications must be made readily available to all users and other interested personnel within the U.S. Navy.

Note to Navy Warfare Library Custodian

This notice should be duplicated for routing to cognizant personnel to keep them informed of changes to this publication.

INTENTIONALLY BLANK

CONTENTS

Page
No.

CHAPTER 1—EVOLUTION OF NAVAL SPECIAL WARFARE

1.1	INTRODUCTION	1-1
1.2	NAVAL COMBAT DEMOLITION UNITS	1-1
1.3	UNDERWATER DEMOLITION TEAMS	1-2
1.4	SEA-AIR-LAND TEAMS	1-3
1.5	SPECIAL BOAT TEAMS	1-5
1.6	SEAL DELIVERY VEHICLE TEAMS	1-5
1.7	UNITED STATES SPECIAL OPERATIONS COMMAND ESTABLISHED	1-5
1.8	NAVAL SPECIAL WARFARE FORCE REALIGNMENT	1-6
1.9	AFGHANISTAN AND IRAQ	1-7
1.10	GROWTH OF NAVAL SPECIAL WARFARE	1-7

CHAPTER 2—NAVAL SPECIAL WARFARE FORCES AND ORGANIZATION

2.1	NAVAL SPECIAL WARFARE FORCES	2-1
2.2	NAVAL SPECIAL WARFARE COMMAND	2-1
2.3	NAVAL SPECIAL WARFARE GROUPS	2-1
2.4	NAVAL SPECIAL WARFARE GROUPS ONE AND TWO	2-2
2.4.1	SEAL Teams	2-3
2.4.2	Logistics Support Units ONE and TWO	2-3
2.4.3	Mobile Communications Team	2-4
2.5	NAVAL SPECIAL WARFARE GROUP THREE	2-5
2.5.1	SEAL Delivery Vehicle Team ONE	2-5
2.5.2	Logistic Support Unit THREE	2-6
2.5.3	Training Detachment THREE	2-7
2.6	NAVAL SPECIAL WARFARE GROUP FOUR	2-7
2.6.1	Special Boat Teams	2-7
2.6.2	Naval Small Craft Instruction and Technical Training School	2-7
2.7	NAVAL SPECIAL WARFARE GROUP TEN	2-8

		Page No.
2.7.1	Support Activity ONE and TWO	2-8
2.7.2	Naval Special Warfare Mission Support Center	2-9
2.7.3	Cultural Engagement Unit	2-9
2.7.4	Naval Special Warfare Anchor Teams	2-9
2.8	NAVAL SPECIAL WARFARE GROUP ELEVEN	2-9
2.9	NAVAL SPECIAL WARFARE UNITS	2-12
2.10	NAVAL SPECIAL WARFARE CENTER	2-12
2.10.1	Basic Training Command	2-12
2.10.2	Advanced Training Command	2-12
2.11	NAVAL SPECIAL WARFARE DEVELOPMENT GROUP	2-12
2.12	CENTER FOR SEAL TEAM AND SPECIAL WARFARE COMBATANT–CRAFT CREWMAN	2-12

CHAPTER 3—COMMAND AND CONTROL

3.1	GLOBAL FORCE MANAGEMENT	3-1
3.1.1	Persistent and Episodic Requirements	3-1
3.1.2	Joint Chiefs of Staff Exercises	3-2
3.1.3	Contingency Plan Execution	3-2
3.1.4	Counterterrorism Execute Order	3-3
3.2	DEFINITIONS	3-3
3.2.1	Assigned Forces	3-3
3.2.2	Attached Forces	3-3
3.2.3	Allocated Forces	3-3
3.2.4	Geographic Combatant Commander Authority	3-3
3.2.5	United States Special Operations Command Authority	3-3
3.2.6	Combatant Command	3-4
3.2.7	Operational Control	3-4
3.2.8	Tactical Control	3-4
3.2.9	Administrative Control	3-4
3.2.10	Support Relationships	3-5
3.2.11	Coordinating Authority	3-5
3.2.12	Direct Liaison Authorized	3-5
3.3	NAVAL SPECIAL WARFARE DEPLOYMENT ROTATION	3-5
3.4	NAVAL SPECIAL WARFARE TASK ORGANIZATION CHAIN OF COMMAND	3-5
3.5	THEATER SECURITY COOPERATION EVENTS	3-7
3.6	NAVAL SPECIAL WARFARE TASK ORGANIZATION COMPONENTS	3-7
3.7	COMMAND AND CONTROL	3-7
3.7.1	Naval Special Warfare Task Force	3-7
3.7.2	Naval Special Warfare Task Group	3-9
3.7.3	Naval Special Warfare Task Unit	3-9

Page No.

3.7.4	SEAL Delivery Vehicle Task Unit	3-9
3.7.5	Special Boat Task Element	3-9
3.7.6	Naval Special Warfare Task Element	3-9
3.8	COMBAT SUPPORT AND COMBAT SERVICE SUPPORT	3-9
3.8.1	Headquarters Support	3-11
3.8.2	Indirect Support	3-11
3.8.3	Direct Support	3-11
3.8.4	Undersea Combat Service Support	3-11
3.8.5	Service-provided Support	3-11
3.9	COMBAT ELEMENTS	3-12
3.10	OTHER DEPLOYABLE NAVAL SPECIAL WARFARE FORCES	3-12
3.11	GEOGRAPHIC COMBATANT COMMANDER ORGANIZATION	3-12
3.11.1	United States Pacific Command	3-13
3.11.2	United States European Command	3-13
3.11.3	United States Central Command	3-14
3.11.4	United States Southern Command	3-14
3.11.5	United States Africa Command	3-16

CHAPTER 4—NAVAL SPECIAL WARFARE CHARACTERISTICS AND EMPLOYMENT

4.1	STRATEGIC CONTEXT FOR EMPLOYMENT	4-1
4.2	NAVAL SPECIAL WARFARE OPERATORS	4-2
4.2.1	Navy SEAL Operator	4-2
4.2.2	SEAL Platoon	4-3
4.2.3	SEAL Delivery Vehicles	4-3
4.2.4	Special Warfare Combatant Craft–Crewman	4-3
4.2.5	Versatility	4-4
4.2.6	Multi-purpose Canine	4-4
4.3	DIRECT ACTION ORIENTATION	4-4
4.4	MARITIME FOCUS	4-5
4.4.1	Seabasing	4-5
4.4.2	Sea-based Command and Control	4-6
4.4.3	Afloat Forward Staging Base	4-6
4.5	TARGET ACCESS	4-6
4.5.1	Intelligence	4-6
4.5.2	Tactical Mobility	4-8
4.6	OTHER SUPPORT	4-8
4.6.1	Explosive Ordnance Disposal Platoon	4-8
4.6.2	Fires	4-8
4.6.3	Shore Basing	4-8
4.7	COMBINED ACTION	4-9
4.7.1	Combined Action Programs	4-9

		Page No.
4.7.2	Joint Combined Exchange Training	4-9
4.7.3	Counternarco-Terrorism Training	4-10
4.7.4	Regional Specialists	4-10
4.7.5	Small Craft Training	4-10
4.8	NAVAL SPECIAL WARFARE TASKS	4-11
4.8.1	Direct Action	4-11
4.8.2	Special Reconnaissance	4-11
4.8.3	Counterinsurgency	4-11
4.8.4	Counterterrorism	4-12
4.8.5	Preparation of the Environment	4-12
4.8.6	Security Force Assistance	4-12
4.8.7	Foreign Internal Defense	4-12
4.8.8	Information Operations	4-12
4.8.9	Unconventional Warfare	4-12
4.9	GUIDELINES FOR EMPLOYMENT	4-13
4.10	LIMITATIONS	4-14

APPENDIX A—TACTICAL MOBILITY

A.1	NAVAL SPECIAL WARFARE TACTICAL MOBILITY	A-1
A.2	SUBSURFACE	A-1
A.2.1	Dry Deck Shelter	A-1
A.2.2	SEAL Delivery Vehicle	A-3
A.3	SURFACE	A-4
A.3.1	Special Boat Detachments	A-4
A.3.2	Combatant Craft–Heavy	A-4
A.3.3	Combatant Craft–Medium	A-4
A.3.4	Combatant Craft–Light	A-7
A.3.5	Master Mariner Program	A-7
A.4	AIR INSERTION	A-7
A.4.1	Maritime Craft Aerial Delivery System	A-7
A.4.2	Rotary Wing	A-7
A.5	TACTICAL GROUND MOBILITY	A-9

LIST OF ILLUSTRATIONS

Page No.

CHAPTER 2—NAVAL SPECIAL WARFARE FORCES AND ORGANIZATION

Figure 2-1.	Naval Special Warfare Chain of Command	2-2
Figure 2-2.	SEAL Team (Active Duty)	2-3
Figure 2-3.	Mobile Communications Team	2-5
Figure 2-4.	SEAL Delivery Vehicle Team	2-6
Figure 2-5.	Special Boat Teams TWELVE and TWENTY	2-8
Figure 2-6.	Special Boat Team TWENTY-TWO	2-8
Figure 2-7.	Naval Special Warfare Group TEN	2-10
Figure 2-8.	Cultural Engagement Unit	2-10
Figure 2-9.	SEAL Teams SEVENTEEN and EIGHTEEN	2-11

CHAPTER 3—COMMAND AND CONTROL

Figure 3-1.	Naval Special Warfare Task Force Structure Example	3-6
Figure 3-2.	Naval Special Warfare Task Organization Components	3-8
Figure 3-3.	Sample Naval Special Warfare Task Unit/Group/Force	3-10
Figure 3-4.	Naval Special Warfare Task Organization in Pacific Command	3-13
Figure 3-5.	Naval Special Warfare Task Organization in European Command	3-14
Figure 3-6.	Naval Special Warfare Task Organization in Central Command	3-15
Figure 3-7.	Naval Special Warfare Task Organization in Southern Command	3-15
Figure 3-8.	Naval Special Warfare Task Organization in Africa Command	3-16

APPENDIX A—TACTICAL MOBILITY

Figure A-1.	Dual and Single Dry Deck Shelter	A-2
Figure A-2.	SEAL Delivery Vehicle Launch from Dry Deck Shelter	A-3
Figure A-3.	Sea Lion	A-5
Figure A-4.	11-meter Rigid Inflatable Boat	A-5
Figure A-5.	Combatant Craft–Assault	A-6
Figure A-6.	Special Operations Craft–Riverine	A-6
Figure A-7.	Kangaroo Duck	A-8
Figure A-8.	Maritime Craft Aerial Delivery System	A-8
Figure A-9.	H-60 Fast Rope Insertion	A-9
Figure A-10.	Light Tactical All-terrain Vehicle	A-10
Figure A-11.	Internally Transportable Vehicle	A-11
Figure A-12.	RG31 A3 Mine-resistant Ambush-protected Vehicle	A-11

INTENTIONALLY BLANK

NWP 3-05

PREFACE

NWP 3-05 (MAY 2013), NAVAL SPECIAL WARFARE, is intended for use by military planners and staff officers and contains information on the nature, forces, organization, and employment of naval special warfare forces.

Throughout this publication, references to other publications imply the effective edition.

Report administrative discrepancies by letter, message, or e-mail to:

COMMANDER
NAVY WARFARE DEVELOPMENT COMMAND
ATTN: DOCTRINE
1528 PIERSEY STREET BLDG O-27
NORFOLK VA 23511-2723

NWDC_NRFK_FLEETPUBS@NAVY.MIL

ORDERING PRINTED COPIES

Order printed copies of a publication using the print-on-demand (POD) system. A command may requisition a publication using the standard military standard requisitioning and issue procedure (MILSTRIP) process on the Naval Supply Systems Command Web site called the Naval Logistics Library (https://nll.ahf.nmci.navy.mil). An approved requisition is forwarded to the specific Defense Logistics Agency (DLA) site at which the publication's electronic file is officially stored. Commands may also order publications through the Navy Doctrine Library System Web site (https://ndls.nwdc.navy.mil) by visiting publication-specific metadata Web pages and selecting the hyperlink on the stock number, which is linked to the Naval Logistics Library Web site. Users may be prompted to create an account to complete the ordering process. Currently, three copies are printed at no cost to the requester.

CHANGE RECOMMENDATIONS

Procedures for recommending changes are provided below.

WEB-BASED CHANGE RECOMMENDATIONS

Recommended changes to this publication may be submitted to the Navy Doctrine Library System, accessible through the Navy Warfare Development Command (NWDC) Web site at: http://ndls.nwdc.navy.smil.mil or https://ndls.nwdc.navy.mil.

URGENT CHANGE RECOMMENDATIONS

When items for changes are considered urgent, send this information by message to the primary review authority, info NWDC. Clearly identify and justify both the proposed change and its urgency. Information addressees should comment as appropriate. See the sample for urgent change recommendation message format on page 15.

ROUTINE CHANGE RECOMMENDATIONS

Submit routine recommended changes to this publication at any time by using the routine change recommendation letter format on page 16. Mail it to the address below or post the recommendation on the Navy Doctrine Library System site.

COMMANDER
NAVY WARFARE DEVELOPMENT COMMAND
ATTN: DOCTRINE
1528 PIERSEY STREET BLDG O-27
NORFOLK VA 23511-2723

CHANGE BARS

Revised text is indicated by a black vertical line in the outside margin of the page, like the one printed next to this paragraph. The change bar indicates added or restated information. A change bar in the margin adjacent to the chapter number and title indicates a new or completely revised chapter.

WARNINGS, CAUTIONS, AND NOTES

The following definitions apply to warnings, cautions, and notes used in this manual:

WARNING

An operating procedure, practice, or condition that may result in injury or death if not carefully observed or followed.

CAUTION

An operating procedure, practice, or condition that may result in damage to equipment if not carefully observed or followed.

Note

An operating procedure, practice, or condition that requires emphasis.

WORDING

Word usage and intended meaning throughout this publication are as follows:

"Shall" indicates the application of a procedure is mandatory.

"Should" indicates the application of a procedure is recommended.

"May" and "need not" indicate the application of a procedure is optional.

"Will" indicates future time. It never indicates any degree of requirement for application of a procedure.

NWP 3-05

```
FM ORIGINATOR
TO (Primary Review Authority)//JJJ//
INFO COMNAVWARDEVCOM NORFOLK VA//
COMUSFLTFORCOM NORFOLK VA//JJJ//
COMUSPACFLT PEARL HARBOR HI//JJJ//
(Additional Commands as Appropriate)//JJJ//
BT
CLASSIFICATION//N03510//
MSGID/GENADMIN/(Organization ID)//
SUBJ/URGENT CHANGE RECOMMENDATION FOR (Publication Short Title)//
REF/A/DOC/NTTP 1-01//
POC/(Command Representative)//
RMKS/ 1. IAW REF A URGENT CHANGE IS RECOMMENDED FOR (Publication Short Title)
2. PAGE _____ ART/PARA NO _____ LINE NO _____ FIG NO _____
3. PROPOSED NEW TEXT (Include classification)




4. JUSTIFICATION.
BT
```

Ensure that actual message conforms to MTF requirements.

Urgent Change Recommendation Message Format

NWP 3-05

DEPARTMENT OF THE NAVY
NAME OF ACTIVITY
STREET ADDRESS
CITY, STATE XXXXX-XXXX

5219
Code/Serial
Date

FROM: *(Name, Grade or Title, Activity, Location)*
TO: *(Primary Review Authority)*

SUBJECT: ROUTINE CHANGE RECOMMENDATION TO *(Publication Short Title, Revision/Edition, Change Number, Publication Long Title)*

ENCL: *(List Attached Tables, Figures, etc.)*

1. The following changes are recommended for NTTP X-XX, Rev. X, Change X:

 a. CHANGE: (Page 1-1, Paragraph 1.1.1, Line 1)
Replace "…the ~~National Command Authority~~ President and Secretary of Defense establishes procedures for the…"
REASON: SECNAVINST ####, dated ####, instructing the term "National Command Authority" be replaced with "President and Secretary of Defense."

 b. ADD: (Page 2-1, Paragraph 2.2, Line 4)
Add sentence at end of paragraph "See Figure 2-1."
REASON: Sentence will refer reader to enclosed illustration.
Add Figure 2-1 (see enclosure) where appropriate.
REASON: Enclosed figure helps clarify text in Paragraph 2.2.

 c. DELETE: (Page 4-2, Paragraph 4.2.2, Line 3)
Remove "Navy Tactical Support Activity."
"…~~Navy Tactical Support Activity, and~~ the Navy Warfare Development Command ~~are~~ is responsible for…"
REASON: Activity has been deactivated.

2. Point of contact for this action is *(name, grade or title, telephone, e-mail address)*.

(SIGNATURE)
NAME

Copy to:
COMUSFLTFORCOM
COMUSPACFLT
COMNAVWARDEVCOM

Routine Change Recommendation Letter Format

NWP 3-05

EXECUTIVE SUMMARY

This navy warfare publication describes the nature, forces, organization, and employment of naval special warfare (NSW).

Naval special warfare is a relatively small, maritime special operations force (SOF) consisting of approximately 9,250 personnel: 2,700 sea-air-land commandos (SEALS), 700 special warfare combatant-craft crewmen (SWCCs), 750 Reservists, 4,000 combat support (CS) and combat service support (CSS) personnel, and more than 1,100 civilians. NSW constitutes 11 percent of special operations forces and less than 2 percent of United States Navy (USN) forces. NSW costs 3/10ths of one cent (0.3 percent) of every United States (U.S.) defense dollar.

Naval special warfare origins go back to World War II (WWII) naval combat demolition units (NCDUs), underwater demolition teams (UDTs), and other special-purpose maritime organizations. In 1962, the first SEAL teams were established. Since 9/11, NSW has grown capacity and added substantial capability, including intelligence gathering and analysis, logistical support, and tactical mobility.

The current NSW organization is based around eight SEAL teams, one SEAL delivery vehicle team (SDVT), three special boat teams (SBTs), and several supporting organizations. These teams and supporting commands are organized into six NSW groups, a training command, and a tactical development group.

NSW is a versatile, maritime-focused, direct action (DA)-oriented force that emphasizes innovation, stealth, precision, and aggressive action. NSW capabilities are based on a rigorous all-volunteer training and selection process stressing mental and physical toughness for SEAL and SWCC operators. NSW tactical mobility assets include combatant craft, undersea combat submersibles, and tactical ground vehicles. Naval special warfare forces also include intelligence, surveillance, and reconnaissance (ISR); cultural engagement; and logistics support units (LOGSUs). These forces combine into scaled and tailored force packages that integrate intelligence gathering and analysis, a versatile portfolio of insertion and extraction techniques, flexible command and control (C2) options, and a strike capability characterized by attention to detail, stealth, speed, precision, and a high density of specialized skills.

NSW is an extreme-environment force operating deep at sea; through the surf; from the air; in jungles, deserts, or high mountains; and in all weather, day or night. NSW operates in all terrain but is uniquely trained and equipped to operate in maritime and riverine areas. The ability to operate over, under, on, and from the sea—as well as in the air and on land—allows access to targets that may be inaccessible to other forces and is whence the acronym SEAL originates.

Naval special warfare forces are employed in all scenarios from peacetime to conventional war. NSW capabilities are particularly well-suited to irregular warfare conflicts involving violent extremist organizations, counterinsurgencies, or other scenarios in which the U.S. interests are threatened by widely dispersed, hard-to-locate, hard-to-get-to, high-priority targets that require precision assaults. NSW forces should be employed when target characteristics include, or target engagement is facilitated by, one or more of the following.

1. Access. Target is hard to locate, hard to get to, or requires collaboration with a partner force.

2. Maritime or riverine.

3. Stealth. Small U.S. signature for political purposes and small tactical signature for military purposes.

4. Precision. Eyes-on, hands-on target discrimination and precise, measured application of force.

NWP 3-05

While this direct approach will always be necessary and is a fundamental part of NSW operations, most security challenges also require the indirect application of military power to build partner capacity and to develop mutually supporting relationships. NSW's contribution to this indirect approach is an intelligence-intensive, human-focused, regionally oriented, persistent presence in key locations throughout the world. These low-cost, small-footprint operations are the norm for most deployed NSW forces and are ongoing daily throughout the world.

NWP 3-05

CHAPTER 1
Evolution of Naval Special Warfare

1.1 INTRODUCTION

Today's naval special warfare personnel can trace their origins back to the amphibious scouts and raiders (S&Rs), NCDUs, Office of Strategic Services Maritime Unit, and UDTs of World War II. While none of those early organizations have survived to the present, their pioneering efforts are mirrored in the missions and professionalism of present NSW operators.

1.2 NAVAL COMBAT DEMOLITION UNITS

In September 1942, 17 Navy salvage personnel arrived at Naval Amphibious Base Little Creek, VA, for a 1-week course on demolitions, explosive cable cutting, and commando raiding techniques. November 10, 1942, this first combat demolition unit succeeded in cutting a cable and net barrier across the Wadi Sebou River during Operation TORCH in north Africa. The unit's actions enabled the USS *Dallas* (DD 199) to traverse the river and insert United States Army Rangers to capture the Port Lyautey airdrome. During the same period at Little Creek, the amphibious S&R (joint) were formed to participate in Operation TORCH to lead amphibious assault forces to the beaches of north Africa. One of the S&R operators was Chief Petty Officer (later Captain) Phil Bucklew (for whom the Naval Special Warfare Training Center is named).

May 14, 1943, the first phase of an NCDU project was established at Solomons, MD, to train six officers and 18 enlisted volunteers to participate in Operation HUSKY, the invasion of southern Italy. The second phase of the project began with planning for a massive cross-channel invasion of Europe after intelligence indicated German forces were arraying a barrier to invasion of underwater obstacles on the beaches at Normandy.

June 6, 1943, Lieutenant Commander Draper L. Kauffman, "the father of naval combat demolition", established the Naval Combat Demolition Unit Training School at Fort Pierce, FL, to train personnel specifically for Europe, although others were deployed to the Pacific. The NCDUs, composed of six-man teams, were formed with volunteers acquired from the Navy Construction Battalion (Seabee) training school at Camp Peary, Williamsburg, VA. During November 1943, NCDU–2, under Lieutenant Junior Grade (LTJG) Frank Kaine (for whom the Naval Special Warfare Command (NAVSPECWARCOM) building is named), and NCDU–3, under LTJG Lloyd Anderson, formed the nucleus of six NCDUs that served with the Seventh Amphibious Force from Biak to Borneo, clearing boat channels after amphibious landings.

By April 1944, 34 NCDUs were deployed to England in preparation for Operation OVERLORD, the Allied invasion of France on the beaches of Normandy. June 6, 1944, in the face of great adversity the NCDUs at Omaha Beach managed to open eight complete and two partial gaps in the German defenses. The NCDUs suffered 31 dead and 60 wounded-a casualty rate of 52 percent. The NCDUs at Utah Beach met less intense enemy fire, clearing 700 yards of beach in 2 hours and clearing another 900 yards by the end of the day. Casualties at Utah Beach totaled six dead and 11 wounded. Not a single demolitioneer was lost to the improper handling of explosives during OVERLORD.

> Professional knowledge has a history, and some knowledge of that history is essential to professional competence.
>
> —*The Soldier and the State*, Samuel P. Huntington

NWP 3-05

In August 1944, the NCDUs that operated at Utah Beach participated in the amphibious landings in southern France, the last amphibious operation in the European Theater of Operations. The majority of NCDUs were then transferred to the Pacific Theater of Operations and eventually absorbed into the UDTs.

The NCDUs at Omaha Beach were presented one of only three Presidential Unit Citations awarded for the Normandy invasion, and the NCDUs at Utah Beach were presented the only Navy Unit Commendation awarded for D-day.

1.3 UNDERWATER DEMOLITION TEAMS

Admiral Chester Nimitz's "Granite Plan" for central Pacific Theater operations required a robust amphibious force. Many of the targeted islands were formed of coral atolls with reefs that served as an obstacle to amphibious assaults. November 23, 1943, problems with the United States Marine Corps (USMC) landing on Tarawa Atoll made clear the need for hydrographic reconnaissance and underwater demolition of obstacles prior to the conduct of amphibious landings. In response to this need, 30 officers and 150 enlisted volunteers were moved to Waimanalo Amphibious Training Base, Oahu, HI, to form the nucleus of what later became UDT-1 and UDT-2.

UDTs first saw combat January 31, 1944, during Operation FLINTLOCK in the Marshall Islands. FLINTLOCK was the catalyst for the formal establishment of a UDT training program in the Pacific Theater. In February 1944, the Naval Combat Demolition Training and Experimental Base was established next to the Amphibious Base at Kamaole at Kihei, Maui, HI.

A total of 34 UDTs were eventually established. Wearing swimsuits, fins, and face masks during combat operations, these "naked warriors" saw action across the Pacific Theater in every major amphibious assault, including Eniwetok, Saipan, Guam, Tinian, Angaur, Ulithi, Pelilui, Leyte, Lingayen Gulf, Zambales, Iwo Jima, Okinawa, Labuan, Brunei Bay, and the July 4, 1945 operation in Balikpapan, Borneo, which was the last major UDT operation of the war.

During the post-war period, the number of active duty UDTs was reduced to two on each coast with a complement of seven officers and 45 enlisted men each. The senior UDT commanding officer (CO) on each coast was also designated as commander, underwater demolition unit (COMUDU), or COMUDU ONE (COMUDO-1) (Coronado, CA) and COMUDU TWO (COMUDO-2) (Little Creek, VA). COMUDU would be the single voice addressing the respective Atlantic and Pacific amphibious force commanders. During the post-war period, COMUDO-1 included UDT-1 and UDT-3; UDT-5 was established in 1952 to provide rotational relief to the other teams rotating through Korea. COMUDO-2 included UDT-2 and UDT-4.

The Korean War began June 25, 1950. Initially, a detachment of 11 personnel from UDT-3 was committed into action in Korea and later expanded to three teams of approximately 100 men each rotating into the Korean Theater of Operations. UDTs, operating under the special operations group, conducted demolition raids against railroad tunnels, bridges, and other targets along the Korean coast.

September 15, 1950, UDTs supported Operation CHROMITE, the amphibious landing at Inchon, Korea. UDT-1 and UDT-3 preceded the amphibious assault, scouting mud flats, marking low points in the channel, clearing fouled assault craft propellers, and searching for mines. Four UDT personnel then acted as wave guides for the subsequent Navy-Marine amphibious assault.

In October, 1950, UDTs supported mine-clearing operations in Wonsan Harbor, Korea, locating and marking mines for clearance by minesweepers. October 12, 1950, UDTs rescued 25 sailors after two U.S. minesweepers struck mines and sank. The next day, UDT operator William Giannotti conducted the first U.S. combat operation using an aqualung when he dove to the sunken minesweeper USS *Pledge* to recover material. Additionally, UDTs conducted beach and river reconnaissance, mine-sweeping operations, infiltrated guerrillas into enemy-held territory from the sea, and participated in Operation FISHNET to cripple the North Korean fishing fleet.

After Korea, UDT-2 and UDT-4 on the east coast were redesignated UDT-21 and UDT-22. The west coast teams became UDT-11, UDT-12, and UDT-13. UDT-13 was disestablished in 1954 and UDT-22 in 1957. UDT-22 would be reestablished when the SEAL teams were authorized in 1961.

The UDTs again saw combat in Vietnam supporting amphibious ready groups and, when attached to riverine patrol groups, they conducted beach and river reconnaissance, destroyed Viet Cong bunkers, accomplished numerous canal clearance operations, and conducted counterinsurgency (COIN) operations working from U.S. and Vietnamese patrol boats. UDT personnel also served as advisors to the South Vietnamese navy.

In a series of 1970's reorganizations, UDT-22 and UDT-13 were established and eliminated again.

1.4 SEA-AIR-LAND TEAMS

A June 5, 1961 letter from the Chief of Naval Operations (CNO) to the commanders of the United States Atlantic Fleet and United States Pacific Fleet outlined "concepts for naval operations in restricted waters, rivers, and estuaries…[and] with particular reference to the conduct and support of paramilitary operations, it is desirable to establish special operations teams as a separate component within Underwater Demolition Units ONE and TWO. An appropriate cover name for such units is 'SEAL' units, being a contraction of SEA, AIR, and LAND."[1]

A CNO letter of October 31 read: "As a first step in improving naval unconventional warfare (UW) capabilities, efforts are being exerted to obtain personnel and funds to activate the two SEAL Units for participation in the water-borne aspects of unconventional warfare."[2] An interesting aspect of this letter is that the narrative used the term "naval special warfare"; this may have been the first, if not one of the earliest, use of the phrase.

SEAL teams were officially authorized in a December 11, 1961 CNO letter, the same letter that reestablished UDT-22. SEAL Teams ONE and TWO, established in January, 1962, were formed entirely of personnel transferred from UDTs.

The SEAL or NSW breast insignia was approved October 16, 1970. When first established, there were separate UDT and SEAL insignia in both silver (enlisted) and gold (officer) versions; however, within several years it was reduced to the existing and enduring gold trident. SEAL enlisted men were authorized to wear a gold insignia because of the equality of warfare area qualification training found only in naval special warfare, gold insignia are generally reserved only for naval officers.

Naval Operation Support Groups Atlantic and Pacific were established in the Atlantic and Pacific October 10, 1963, and were the predecessor of today's naval special warfare groups (NSWGs). They supplanted COMUDU as senior in the chain of command.

SEAL involvement in Vietnam began initially as advisors to the Vietnamese in the conduct of clandestine maritime operations. SEALs also began a UDT-style training course for the Biet Hai (sea commandos) based in Danang, Vietnam.

In February 1966, a small SEAL Team ONE detachment arrived in Vietnam to conduct direct-action missions out of Nha Be in the Rung Sat Special Zone. This deployment would evolve into a steady state presence of eight SEAL platoons for a number of years. Additionally, SEALs served as advisors for provincial reconnaissance units and the Lien Doc Nguoi Nhia, the Vietnamese SEALs. Although small in size, SEALs were among the most highly decorated units in the Vietnam War. Decorations included three Medals of Honor and five Navy Crosses. The last SEAL platoon left Vietnam December, 7, 1971, and the last SEAL advisor left in March 1973.

NSW was reorganized several times during the 1970s, the size of the SEAL teams was dramatically reduced, and the NSWGs were briefly replaced by other organizations.

---

[1] CNO letter, June 5, 1961.
[2] CNO letter, October 31, 1961.

From 1982 thru 1993, NSW personnel served as trainers and advisors in El Salvador and maintained a small advisor base in La Union, El Salvador.

May 1, 1983, UDT-11 was redesignated as SEAL Team FIVE, UDT-21 was redesignated as SEAL Team FOUR, UDT-12 was redesignated as SEAL Delivery Vehicle Team ONE (SDVT-1), and UDT-22 was redesignated as SDVT-2. A new team, SEAL Team THREE, was established October 1, 1983 in Coronado, CA.

In 1983, SEALs participated in Operation URGENT FURY, the invasion of Grenada. NSW forces conducted a night beach reconnaissance in preparation for an amphibious assault, conducted a direct action against an enemy controlled radio station, and fought off a large enemy force in a mission to protect the governor general and his family.

SEALs participated in Operation EARNEST WILL in the Persian Gulf from 1987 to 1989. Two oil-servicing barges, the *Hercules* and the *Wimbrown VII*, were converted into mobile sea bases from which NSW forces conducted patrol and interdiction operations in the northern Persian Gulf to maintain the security of sea lines of communication. Missions included interdiction of Iranian mine-laying operations, small-boat harassments, and attacks on shipping, including the capture of the *Iran Ajr*, an Iranian minelayer.

SEAL Team EIGHT was established on October 1, 1988 at Naval Amphibious Base, Little Creek, VA.

In 1989, SEALs participated in Operation JUST CAUSE, the invasion of Panama to topple the Noriega dictatorship. Operating out of Rodman Naval Station, Panama, NSW forces secured the Atlantic and Pacific entrances to the Panama Canal, denied Noriega the use of his aircraft located at Paitilla Airfield, sank the Panamanian patrol boat *Presidente Poras*, and conducted numerous reconnaissance and DA search and seizure missions.

From August 1990 thru March 1991, SEALs participated in Operation DESERT SHIELD and Operation DESERT STORM. They conducted beach and land border reconnaissance missions, 118 combat search and rescue (CSAR) missions, and 92 mine countermeasures missions. SEALs liberated the first Kuwaiti territory, captured the first enemy prisoners of war, and conducted a maritime CSAR operation. SEALs also conducted a maritime deception mission, a feint that successfully drew Iraqi forces away from the point of the U.S. assault into Kuwait.

From December 1992 thru March 1995, NSW forces participated in Operation RESTORE HOPE in Somalia. They supported amphibious operations by performing amphibious and other special reconnaissance, riverine operations, harbor and port surveys, convoy and personnel protection, and sniper operations.

From May thru September 1994, NSW forces (Naval Special Warfare Task Unit (NSWTU)-Wasp) conducted amphibious reconnaissance and other special operations as part of Operation SUPPORT DEMOCRACY and Operation UPHOLD DEMOCRACY in Haiti. In addition to supporting amphibious reconnaissance, NSW patrol coastal ships participated in the enforcement of the United Nations embargo against Haiti by conducting maritime interdiction operations.

From December 1995 thru June 1998, as part of Operation JOINT ENDEAVOR and Operation JOINT GUARD in Bosnia, NSW forces conducted reconnaissance and served as a quick reaction force for the Sava River bridging operations. They also served as joint commission observers in Bosnia, six-man elements that facilitated communications between formerly warring factions and the United Nations Stabilization Force commanders.

From April 7-20, 1996, personnel from Naval Special Warfare Unit TWO (NSWU-2) participated in Operation ASSURED RESPONSE, during which they conducted a noncombatant evacuation operation (NEO) of more than 2,000 people from the U.S. Embassy in Monrovia, Liberia.

From September 21 thru October 10, 1998, NSWU-2 and NSWU-10 forces participated in Operation SHADOW EXPRESS, a noncombatant evacuation operation in Monrovia, Liberia. A political settlement ended the requirement for a full evacuation.

1.5 SPECIAL BOAT TEAMS

Today's special warfare combatant-craft crewmen trace their roots to the motor torpedo boat squadrons of WWII and especially from the Vietnam-era mobile support teams (MSTs) of the NSW boat support units (BSUs). In February 1964, Boat Support Unit ONE (BSU–1) was established under Naval Operations Support Group, Pacific, to operate patrol torpedo fasts (naval attack craft) (PTFs) and other high-speed craft conducting special operations and coastal patrol and interdiction (CP&I). Late in 1964, the first PTFs arrived in Danang, Vietnam. In 1965, BSU–1 began training Vietnamese patrol craft fast (Swift Boat) crews to conduct CP&I. As the Vietnam mission expanded, additional NSW craft, tactics, and training evolved, including insertion, extraction, and fire support for SEAL operations. BSU–1 crews formed into MSTs deployed with and supported every SEAL platoon engaged in Vietnam.

NSW combatant craft units were reorganized twice in the 1970s and again in 1994. In 1994, the organizational structure included Special Boat Units (SBUs) ELEVEN (SBU–11) and TWELVE (SBU–12) under Special Boat Squadron ONE (SBR–1) (west coast), and Special Boat Units TWENTY (SBU–20) and TWENTY TWO (SBU–22) (east coast) under Special Boat Squadron TWO (SBR–2) (east coast). SBU–11 was disestablished in September 1997.

During Operation EARNEST WILL in the Persian Gulf from 1987 thru 1989, SBUs operated 65-foot patrol boat (PB) MK III craft from the *Hercules* and *Wimbrown VII*. The PBs conducted extensive patrol and interdiction operations in the northern Persian Gulf (see section 1.4 above).

April 1, 1994, the SWCC enlisted warfare specialty was established allowing the NSW force to retain highly trained boat crews rather than have them return to the fleet after one tour. In 2001, the SWCC breast insignia was approved, and in October 2006, the Navy established the special warfare boat operator (USN) rating allowing boat crewman to stay in NSW for their entire career.

1.6 SEAL DELIVERY VEHICLE TEAMS

The SDVTs trace their roots to Office of Strategic Services maritime unit swimmer groups and their use of the British submersible canoe, "Sleeping Beauty," introduced to the post-war UDTs in 1947 at St. Thomas, U.S. Virgin Islands. The UDTs aggressively pursued development of swimmer propulsion units and free-flooding SEAL delivery vehicle (SDV) capabilities throughout the 1950s and into the 1960s with the fielding of the MK 7 Mod 0 SDV, the first reliable and deployable combatant submersible in the NSW inventory; MK 8 and 9 SDVs were introduced in the late 1970s. Today, the MK 8 Mod 1 SDV represents the state of the art in special operations underwater mobility.

May 1, 1983, UDT–12 was redesignated as SDVT–1, and UDT–22 was redesignated as SDVT–2. August 8, 2008, SDVT–2 was disestablished, leaving SDVT–1 as the only remaining SDV team. SDVT–1 moved from Coronado, CA, to Pearl Harbor, HI, in 1994.

1.7 UNITED STATES SPECIAL OPERATIONS COMMAND ESTABLISHED

To ensure that special operations forces maintain a high state of readiness and to correct deficiencies accentuated by the failed attempt to rescue American hostages in Iran in April 1980, a comprehensive program of SOF revitalization began in 1981. Congress authorized significant budget increases between 1981 and 1986, but the slow pace of the revitalization, lack of attention to special operations from the parent Services, limited career opportunity for SOF personnel, difficulties in satisfying SOF-unique material requirements, and the absence of any high-level spokesmen for special operations within the Department of Defense (DOD) prompted additional congressional action.

The culmination of bipartisan congressional efforts to improve DOD special operations and low-intensity conflict capability was the Cohen-Nunn Amendment to the 1987 Defense Authorization Act. This law directed establishment of a new unified combatant commander (CCDR) in the commander, United States Special Operations Command (CDRUSSOCOM), and the civilian post of Assistant Secretary of Defense for Special

NWP 3-05

Operations and Low-intensity Conflict within the Office of the Secretary of Defense. This legislation also created Major Force Program 11 of the Future Years Defense Program, which consolidated all SOF resourcing under one DOD Major Force Program budget account.

United States Special Operations Command (USSOCOM) was formally established April 16, 1987, by Congress as a unified command to provide special operations forces the funding and organizational relationships necessary to field a professional U.S. special operations capability. CDRUSSOCOM exercises combatant command (command authority) (COCOM) over all U.S.-based SOF. Operational control (OPCON) and administrative control (ADCON) is exercised primarily through USSOCOM's Service components. The naval component, also established April 16, 1987, is NAVSPECWARCOM. Prior to that time, east and west coast NSW forces had been assigned to their respective naval surface force commanders.

The establishment of USSOCOM gave NSW a dual higher headquarters (HQ) relationship. NSW was assigned to USSOCOM as its naval component but retained strong ties to the Navy as its special operations component. On Service-specific issues, such as personnel, administration, logistics, equipment, and individual training, Commander, United States Special Warfare Command (COMNAVSPECWARCOM) communicates directly with the CNO, keeping CDRUSSOCOM informed.

1.8 NAVAL SPECIAL WARFARE FORCE REALIGNMENT

In 2001, COMNAVSPECWARCOM initiated a realignment/reorganization with the following objectives:

1. Deploy a force with improved capability for C2 of geographically dispersed operations
2. Deploy senior NSW leadership (e.g., SEAL team COs) forward to lead the forces they train and equip
3. Reduce administrative and logistics burdens on SEAL team commanders to allow them to better focus on operations and training.

The strategy consolidated unit training under training detachments (TRADETs), consolidated logistics under LOGSU, and consolidated qualification training under the Naval Special Warfare Center (NSWC). These improvements helped standardize fundamental training and equipping functions. They also consolidated resources, which enabled more rapid adaptations in advanced training such as tactical ground mobility and tactical unmanned aircraft systems (UASs).

This realignment was accomplished over a period of several years. In 2001, NSWGs 1 and 2 and their respective SEAL teams were realigned. Six SEAL teams were reorganized into eight SEAL teams with the establishment of SEAL Team SEVEN, March 17, 2002; and SEAL Team TEN, April 19, 2002. The majority of the old SEAL team administrative and CSS personnel were transferred to the new NSW LOGSUs; the majority of personnel dedicated to training SEAL platoons were transferred to two TRADETs under NSWGs 1 and 2. To ensure common training standards across the force, NSWC took responsibility for providing most individual training courses for NSW.

Six months prior to deployment, a SEAL team is augmented with a Navy special operations forces (NAVSOF) explosive ordnance disposal (EOD) platoon, a combat service support team from a LOGSU, a mobile communications team (MCT), a cross-functional team (intelligence support), and other attachments as required by the anticipated mission taskings. These SEAL teams with CS and CSS personnel are designated as a naval special warfare task group (NSWTG). Deployed NSWTGs support the operational requirements of geographic combatant commanders (GCCs).

To complete the reorganization, October 1, 2002 SBR–1 was redesignated as Naval Special Warfare Group THREE and SBR–2 was redesignated as Naval Special Warfare Group FOUR. The three SBUs were redesignated as special boat teams (SBTs), and SDVTs 1 and 2 were transferred to NSWGs 3 and 4. Two years later, October 1, 2004, subsurface and surface craft responsibilities were split. Commander, Naval Special Warfare Group (CNSWG) THREE (CNSWG–3) was designated the immediate superior in command of both SEAL

NWP 3-05

delivery vehicle teams (consolidated into one team, SDVT-1, in 2008) and CNSWG-4 was designated the immediate superior in command of the three SBTs. (Chapter 2 shows the current organization.)

1.9 AFGHANISTAN AND IRAQ

Immediately following the terrorist attacks of 9/11, NSW forces deployed to Afghanistan for Operation ENDURING FREEDOM. Commander, Naval Special Warfare Group ONE (CNSWG-1), serving as Commander, Combined Joint Special Operations Task Force K-BAR, led a SOF task force (TF) of more than 2,500 personnel, including SOF from several coalition countries. In the first 6 months of ENDURING FREEDOM, SEALs participated in 75 special reconnaissance (SR) and DA missions, including leadership interdiction operations, and destroyed more than 500,000 pounds of captured explosives and weapons.

As the war in Afghanistan progressed, NSW continued operations to interdict Taliban networks with increasing emphasis on combined operations with Afghan National Security Force partners. Starting in 2009, NSW forces from Task Unit Trident in southern Afghanistan conducted village stability operations to provide a persistent presence within the local population and support to the Afghan Local Police.

From March 2003 thru September 2010, SEAL and SWCC forces participated in Operation IRAQI FREEDOM. In March 2003, NAVSOF captured the southern oil facilities of the Al-Faw Peninsula, including off-shore gas and oil terminals. NSW SEAL and SWCC forces cleared the Khor Al Abdullah and Khor Az Zubayar waterways, enabling humanitarian aid delivery to the vital port city of Umm Qasr. SEALs also conducted the first successful U.S. prisoner of war rescue since WWII. As the war progressed, NSW forces conducted many COIN/terrorism operations in Iraq, including DA, SR, foreign internal defense (FID)/combat advisor, and personal security detachment operations for Iraqi civilian government senior leadership.

NSW forces remained in Iraq through 2011, advising and assisting Iraqi security forces during Operation NEW DAWN.

1.10 GROWTH OF NAVAL SPECIAL WARFARE

The end of the Cold War in 1989 and the terrorist attacks on the United States September 11, 2001, prompted growth in all special operations forces and resulted in a more-than-50-percent growth in the size of the NSW force. Between 2001 and 2011, four new captain (O-6)-level commands—NSWGs 10 and 11, NSWU-10, and the Center for SEAL Team and Special Warfare Combatant-Craft Crewmen (CENSEALSWCC)—and 11 new commander (O-5)-level NSW commands were established: SEAL Teams SEVEN and TEN, Reserve SEAL Teams SEVENTEEN and EIGHTEEN, Support Activity ONE (SUPPACT-1) and TWO (SUPPACT-2), the Basic Training Command (BTC) and the Advanced Training Command (ATC), and three new LOGSUs. These changes significantly improved NSW's ability to support commanders overseas.

In 2001, LOGSUs ONE (LOGSU-1) and TWO (LOGSU-2) were created as subordinate commands to NSWGs 1 and 2 to provide logistics support for training and to deploy combat service support teams to support NSW forces overseas. TRADET-3 and LOGSU THREE (LOGSU-3), both subordinate to NSWG-3, were commissioned in January 2010 and 2011, respectively. LOGSU-3 specializes in undersea combat service support (UCSS).

SUPPACTs 1 and 2, established in December 2006 and July 2007, respectively, are tasked to increase the "find and fix" capabilities of the force by conducting preparation of the environment (PE) and ISR. NSWG-10 was commissioned May 25, 2011 to consolidate command of the support activities. NSWG-10 also took command of the NSW mission support center (MSC) and the NSW cultural engagement unit (CEU). The MSC provides a central reachback capability for time-sensitive intelligence analysis to support forward-deployed NSW forces and also serves as a repository of intelligence information for mission planning. The CEU deploys native-born and female enablers to support forward-deployed NSW forces

Operation IRAQI FREEDOM was a catalyst for refinement of the find, fix, finish, exploit, and analyze (F3EA) targeting methodology. At the beginning of the Iraq War, rudimentary techniques yielded sporadic results.

Refinement and professionalization of intelligence synthesis by the support activities during the war improved performance so that a "jackpot" (successful capture or kill of a high-value individual (HVI)) became routine.

In October 2005, CENSEALSWCC was established to support career management, warrior development, and professional military education.

NSWG-11 and its two subordinate commands, SEAL Teams SEVENTEEN and EIGHTEEN, were established August 22, 2008. NSWG-11 commands NSW Reserves.

NSWU-10 was reestablished April 8, 2011 and supports United States Africa Command (USAFRICOM). NSWUs 1, 2, and 3 had been previously established providing forward-based, persistent NSW presence in United States Pacific Command (USPACOM), United States European Command (USEUCOM), and United States Central Command (USCENTCOM).

NSWU-4 was reestablished in 2011 as an officer in charge (OIC) reporting to NSWG-2 and supports United States Southern Command (USSOUTHCOM).

NWP 3-05

CHAPTER 2
Naval Special Warfare Forces and Organization

2.1 NAVAL SPECIAL WARFARE FORCES

This chapter deals with the administrative, in-garrison organization of NSW. Deployed organization is described in chapter 3 and capabilities are discussed in chapter 4. Note that the terms "NSW forces" and "NAVSOF" are used interchangeably. NSW is the traditional Navy term, but NAVSOF is often used in the joint arena and is consistent with the usage in other Services: ARSOF for Army special operations forces and MARSOF for Marine special operations forces.

NSW is a relatively small maritime special operations force consisting of approximately 9,200 personnel: 2,700 SEALs, 700 SWCC, 700 Reserves, 4,000 CS and CSS personnel, and more than 1,100 civilians. NSW constitutes 11 percent of U.S. special operations forces (USSOF) and less than 2 percent of Navy forces.

NSW is organized around eight SEAL teams, one SDVT, three special boat teams, and supporting commands which deploy forces worldwide to meet the requirements of theater commanders. These commands are organized into six naval special warfare groups and a training command, Naval Special Warfare Center (figure 2-1).

2.2 NAVAL SPECIAL WARFARE COMMAND

Commander, Naval Special Warfare Command, a two-star rear admiral headquartered in Coronado, CA, exercises OPCON and ADCON of all active and Reserve NSW forces stationed in the United States. The NAVSPECWARCOM mission is to man, train, equip, deploy, and sustain NSW forces for operations and activities abroad in support of combatant commanders and U.S. national interests.

COMNAVSPECWARCOM is an echelon II command assigned under the combatant command of USSOCOM. As a Navy commander, COMNAVSPECWARCOM reports to the CNO for Navy-specific issues, including funding for Service-common systems, services, equipment, training, munitions, and materiel. Details of the responsibilities and relationships are delineated in a memorandum of agreement between the Navy and USSOCOM (see References section).

2.3 NAVAL SPECIAL WARFARE GROUPS

Deployable NSW forces are assigned to one of the six NSWGs. SEAL teams are assigned to NSWG-1 or NSWG-2, undersea combat submersibles to NSWG-3, surface combatant craft to NSWG-4, intelligence support to NSWG-10, and Reserves to NSWG-11. An NSWG is an echelon III command under the OPCON and ADCON of NAVSPECWARCOM (echelon II). NSWGs exercise OPCON and ADCON of subordinate commands and assigned forces for COMNAVSPECWARCOM. The mission of the NSWG is to organize, train, equip, deploy, and sustain assigned NSW forces. NSWG commanders are Navy SEALs who hold the rank of captain and are accorded the traditional Navy honorific title of commodore.

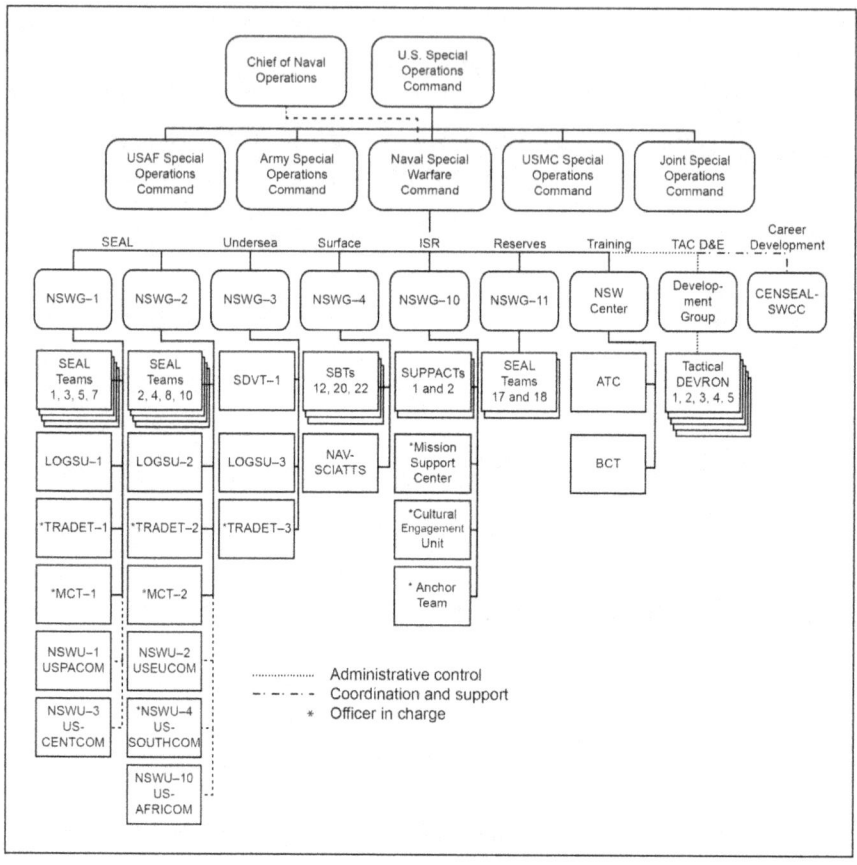

Figure 2-1. Naval Special Warfare Chain of Command

Odd-numbered groups—NSWGs 1, 3, and 11—are stationed at Naval Base Coronado, CA. Even-numbered groups—NSWGs 2, 4, and 10—are stationed at Joint Expeditionary Base Little Creek–Fort Story, VA.

2.4 NAVAL SPECIAL WARFARE GROUPS ONE AND TWO

NSWGs 1 and 2 command the active duty SEAL teams. In addition to preparing and deploying SEALs and support forces, NSWGs 1 and 2 are tasked to be prepared to deploy and function as a naval special warfare task force (NSWTF) or as a joint special operations task force (JSOTF) HQ battle staff. JSOTF is described in Joint Publication (JP) 3-05, Doctrine for Joint Special Operations. NSWGs 1 and 2 also manage the NSW Multi-Purpose Canine program.

NWP 3-05

Commander, Naval Special Warfare Group ONE is designated as the executive agent for NSW support to USCENTCOM and USPACOM. Subordinate commands are SEAL Teams ONE, THREE, FIVE, and SEVEN and LOGSU-1. Subordinate detachments are Mobile Communications Team ONE (MCT-1) and TRADET-1. CNSWG-1 also exercises ADCON of NSWU-1 and NSWU-3.

Commander, Naval Special Warfare Group TWO is designated as the executive agent for NSW support to USAFRICOM, USEUCOM, United States Northern Command (USNORTHCOM), and USSOUTHCOM. Subordinate commands are SEAL Teams TWO, FOUR, EIGHT, and TEN and LOGSU-2. Subordinate detachments are MCT-2 and TRADET-2. CNSWG-2 also exercises ADCON of NSWU-2, NSWU-4, and NSWU-10.

2.4.1 SEAL Teams

There are eight active duty SEAL teams. The SEAL team mission is to organize, man, train, equip, deploy, and employ SEALs and other forces to plan and conduct special operations in support of GCCs. A SEAL team is an echelon IV command subordinate to an NSWG (figure 2-2).

A SEAL team is organized into seven SEAL platoons consisting of three officers and 18 enlisted each. The platoons are organized into three SEAL task units with a small C2 element and two or three SEAL platoons.

2.4.2 Logistics Support Units ONE and TWO

LOGSUs 1 and 2 are subordinate commands of NSWG-1 and NSWG-2. LOGSUs organize, man, train, equip, and deploy personnel and equipment to provide both in-garrison and deployed CSS to SEAL teams and SPECBOAT teams.

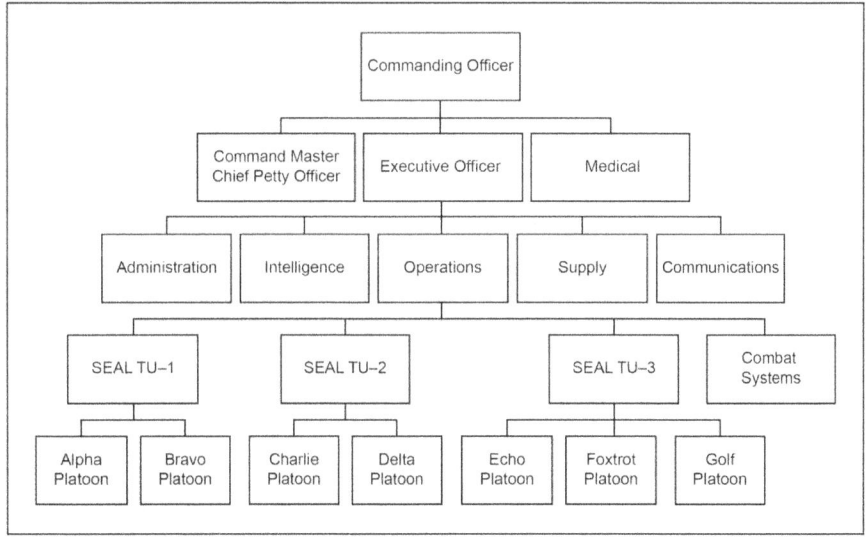

Figure 2-2. SEAL Team (Active Duty)

2-3   MAY 2013

Specific logistics support functions include supply; medical; equipment management and maintenance, including air operations, dive operations, special operations forces personal equipment advanced requirements gear, and weapons systems (weapons, visual acuity systems, and munitions); transportation (vehicle dispatch, line haul, and maintenance/repair); facilities management (base camp, range, and training facility support, including logistics, messing, maintenance, and scheduling); and military construction.

LOGSUs 1 and 2 are organized into an HQ element, administrative departments, and four deployable CSS teams. An 86-person CSS team is led by a lieutenant (O–3) or lieutenant commander (O–4) supply corps officer. Each SEAL team is augmented with a dedicated CSS team as they enter the unit-level training phase of the interdeployment training cycle (IDTC) and receives dedicated supply support throughout the remainder of the IDTC and deployment.

A CSS team can be disaggregated into three CSS detachments. CSS detachments or smaller elements can be assigned to deploy with NSWTUs, special boat detachments (SPECBOATDETs), or other task organizations. Deployed NSW LOGSU CSS teams:

1. Coordinate with the appropriate GCC component commands for support of forward-deployed NSW forces

2. Provide limited contingency contracting capability and expertise, including small purchases and leases, as authorized by the theater executive agent and in coordination with the GCC logistics staff section

3. Conduct force embarkation, transportation, equipment maintenance, combat cargo handling, in-theater logistics, construction, contingency engineering, camp development, and maintenance services

4. Assist with equipment maintenance, facilities management, combat systems support, ordnance management, and hazardous materials handling

5. Provide limited force protection and perimeter defense

6. Assist with chemical, biological, radiological, nuclear, and high-yield explosives decontamination operations.

A CSS detachment may require augmentation to fully perform its mission. In addition to organic CSS, NSW forces generally receive basing support from conventional forces either afloat or ashore. Details are available in NWP 3-05.41, Naval Special Warfare Logistics.

2.4.3 Mobile Communications Team

NSWGs 1 and 2 each have an MCT. The MCT mission is to organize, man, train, equip, and deploy personnel and communications equipment to operate and maintain communications for SEAL teams and SPECBOAT teams.

MCTs provide rapidly deployable mobile communications teams that can be employed in austere environments (figure 2-3). Mobile communications teams deploy with the equipment and skills for a variety of voice and data services for communicating with joint and fleet units. This includes man-portable and modular systems (high frequency, ultrahigh frequency, very high frequency, super high frequency, and extremely high frequency) adequate to support an NSWTG battle staff.

For routine deployments, an MCT is attached to a SEAL team 6 months prior to deployment and supports SEAL platoons and SPECBOATDETs assigned to that team. The MCT can be disaggregated to form multiple mobile communications detachments or elements.

NWP 3-05

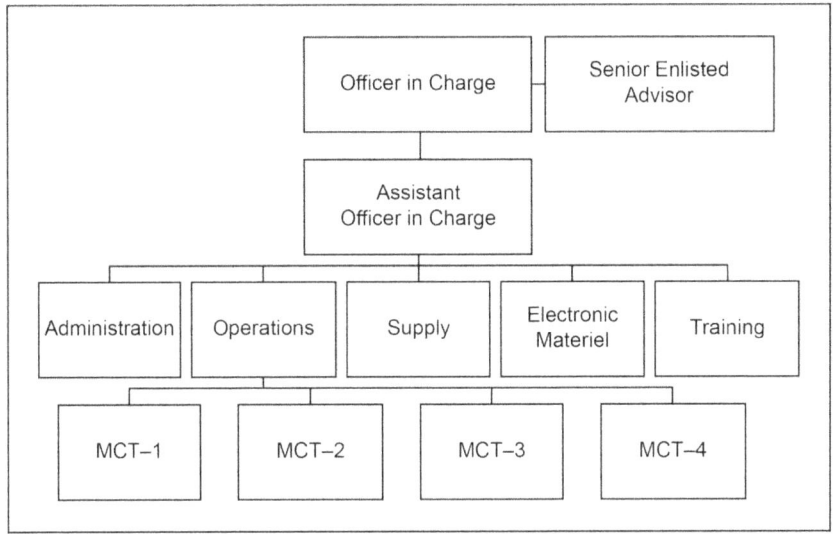

Figure 2-3. Mobile Communications Team

2.5 NAVAL SPECIAL WARFARE GROUP THREE

NSWG-3, as the undersea proponent for NSW, mans, trains, equips, deploys, sustains, and provides command and control for NSW forces conducting undersea special operations worldwide in support of CCDRs and interagency partners. Undersea special operations include activities initiated undersea that continue above the waterline and into the hinterland.

Subordinate commands are SDVT-1, TRADET-3, and LOGSU-3. NSWG-3 also maintains a detachment at Joint Expeditionary Base Little Creek–Fort Story, VA, with an attached dry deck shelter (DDS) platoon. Detachment Little Creek's primary mission is maintenance of and deployment with east coast DDSs.

2.5.1 SEAL Delivery Vehicle Team ONE

SDVT-1, based in Pearl Harbor, HI, conducts operations involving undersea mobility platforms. The command supports national taskings, contingency plan execution, exercises, and other global operations as directed by USSOCOM.

SDVT-1 is organized around five SDV platoons. An SDV platoon is a variation of a SEAL platoon that specializes in SDV operations. It consists of two SDV-qualified SEAL officers, 14 SDV-qualified SEAL enlisted men, and two SDVs. SDV-qualified SEALs have completed the basic SDV school and are certified SDV pilots or navigators.

SDV platoons are organized under SDV task units (SDVTUs). An SDVTU consists of a two-man HQ element and an SDV platoon (figure 2-4). When deploying aboard a DDS-capable submarine, the SDVTU is augmented with a UCSS team and a DDS platoon.

A DDS platoon consists of 26 DDS-qualified Navy divers who operate a DDS. The SDV and DDS are described in appendix A.

2.5.2 Logistic Support Unit THREE

LOGSU-3 has the same mission and functions as LOGSUs 1 and 2 but is unique in that it also has maintenance and material support responsibilities for maritime undersea systems, including SDVs, DDSs, other combat submersibles, and surface support craft. LOGSU-3 deploys UCSS elements to support DDS and SDV operations.

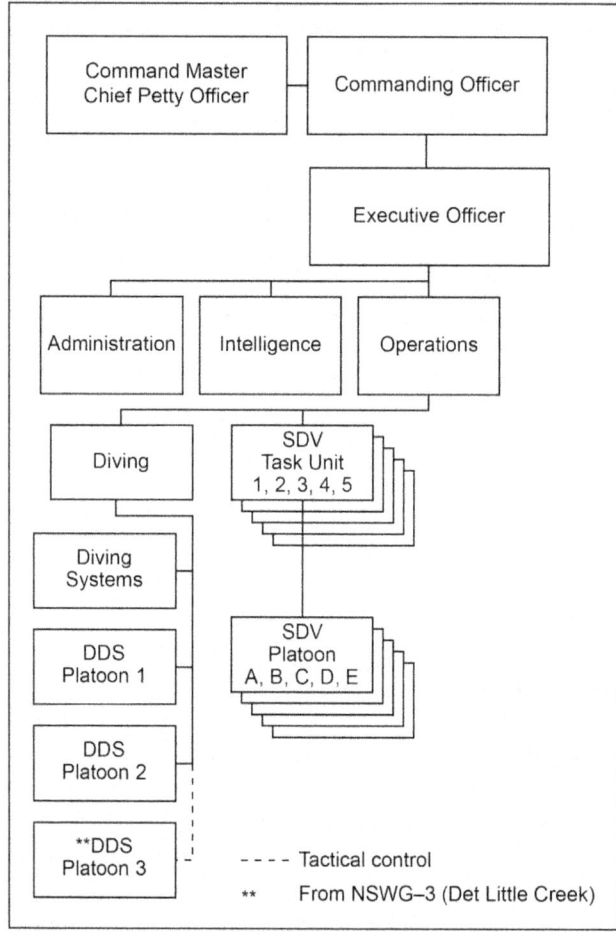

Figure 2-4. SEAL Delivery Vehicle Team

2.5.3 Training Detachment THREE

TRADET-3 has the same function and missions as TRADETs 1 and 2 in training SEAL skill sets but is unique in that it also trains SDV and DDS personnel for undersea missions.

2.6 NAVAL SPECIAL WARFARE GROUP FOUR

NSWG-4—as the surface mobility proponent—organizes, trains, and equips assigned personnel to deploy combat-ready forces and maritime mobility systems in support of fleet and joint commanders and conducts security force assistance (SFA) to build foreign security force small-craft capabilities and capacities in accordance with USSOCOM priorities. Responsibilities include the exercise of operational control and administrative control of Special Boat Teams TWELVE (SBT-12), TWENTY (SBT-20), and TWENTY TWO (SBT-22); and Naval Small Craft Instruction and Technical Training School (NAVSCIATTS); and advising COMNAVSPECWARCOM on combatant craft and other designated combat development areas.

2.6.1 Special Boat Teams

The mission of an SBT is to organize, man, train, equip, and deploy SPECBOATDETs to conduct maritime special operations in riverine, littoral, and open-ocean maritime environments.

NSW special operations craft are categorized as combatant craft-light (CCL), combatant craft-medium (CCM), and combatant craft-heavy (CCH). Craft selection is tailored to the mission and environment. The craft most frequently employed are high-speed combatant craft designed for delivery of SOF in littoral and riverine environments. Lightly armed and armored NSW combatant craft employ stealth, surprise, speed, and agility in the conduct of special operations.

An SBT is organized into an HQ element, administrative and logistics support elements, and deployable SPECBOATDETs. A SPECBOATDET generally consists of two craft, SWCC boat crews, and CSS personnel that accompany the combatant craft for boat maintenance. An exception to the two-craft guideline is a CCH detachment, which may consist of one boat. Boat detachments are organized into teams led by a SEAL O-3 or an SWCC chief warrant officer.

Special Boat Teams TWELVE and TWENTY have 20 and 16 SPECBOATDETs-Coastal, respectively, organized into four special boat teams. Each team consists of four to five SPECBOATDETS-Coastal plus an HQ element (figure 2-5). A SPECBOATDET-Coastal conducts maritime special operations in an open-ocean and littoral environment with a variety of craft. Select coastal detachments are maritime craft aerial delivery system (MCADS)-capable and can be inserted by paradrop. An MCADS detachment is unique in that it has four boats: two boats for routine operations and two boats rigged for paradrop for rapid response. SBTs 12 and 20 also deploy master mariner detachments, four-man detachments trained to operate civilian vessels of any size.

SBT-22 is organized around the special operations craft-riverine (SOC-R). It is organized into four riverine teams with two SPECBOATDETs-Riverine per team. Each SPECBOATDET-Riverine has two SOC-Rs plus an HQ element (figure 2-6). A SPECBOATDET-Riverine conducts maritime special operations in a riverine and littoral environment.

2.6.2 Naval Small Craft Instruction and Technical Training School

The Naval Small Craft Instruction and Technical Training School is an NSWG-4 subordinate command located at the Stennis Space Center in Mississippi.

The mission of NAVSCIATTS is to conduct SFA and foreign internal defense in support of CCDRs using mobile training teams (MTTs) and in-resident training to prepare partner nation forces to conduct small-craft operations in riverine or littoral environments.

2.7 NAVAL SPECIAL WARFARE GROUP TEN

NSWG-10 organizes, trains, equips, deploys, and sustains specialized ISR and PE capabilities through subordinate commands, units, and detachments: SUPPACTs 1 and 2, the mission support center, the NSW cultural engagement unit, and NSW anchor teams.

2.7.1 Support Activity ONE and TWO

SUPPACTs 1 and 2 man, train, equip, organize, and deploy forces to conduct ISR, PE, special technical operations, and combat support in support of NSW forces and GCCs.

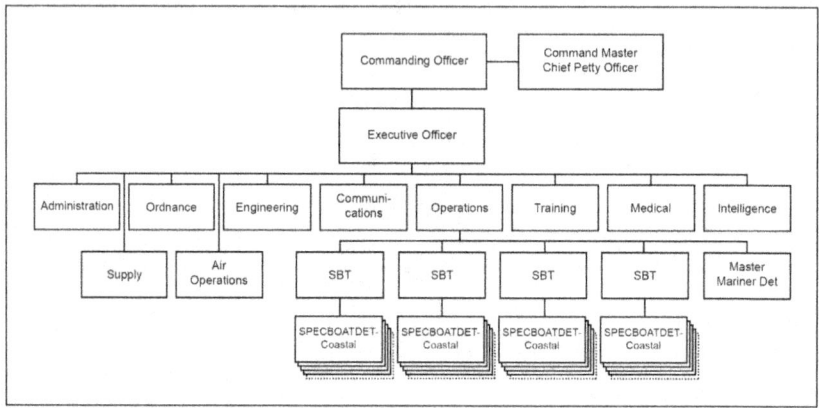

Figure 2-5. Special Boat Teams TWELVE and TWENTY

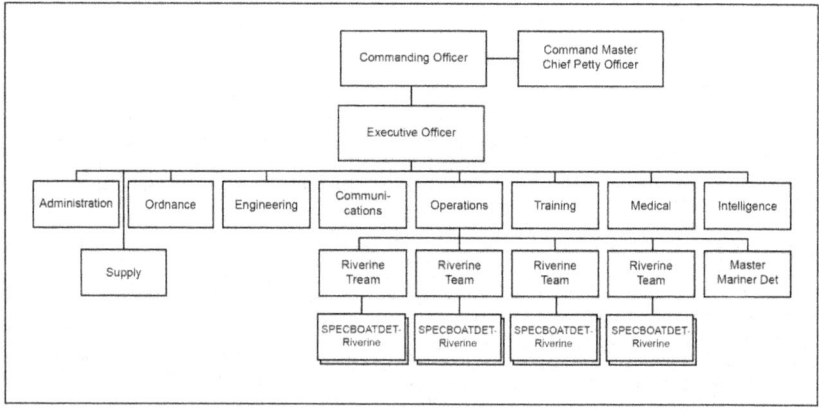

Figure 2-6. Special Boat Team TWENTY-TWO

This capability is provided through cross-functional teams, so named because they combine capabilities from a cross-section of intelligence and operational PE. Cross-functional teams are focused on intelligence collection and analysis to provide support for the find, fix, finish, exploit, analyze, and disseminate (F3EAD) targeting cycle. Personnel come from a cross-section of special operations (SEAL and SWCC) and Navy Information Dominance Corps personnel (intelligence, meteorological and oceanographic (METOC), cryptologists, and electronics and information technicians).

The cross-functional team is organized into a five-person C2 element, a UAS detachment, a METOC detachment, and three cross-functional detachments of two squads each. Each 10-man cross-functional squad includes a human intelligence operations element, analysis and targeting element, tactical information operations (TIO) element, and a technical surveillance element. Cross-functional team personnel are cross-trained in multiple special operations and intelligence disciplines (figure 2-7).

To best satisfy mission requirements, cross-functional teams are often disaggregated into small task-organized elements with the most appropriate mix of skill sets for a given deployment or operation.

2.7.2 Naval Special Warfare Mission Support Center

The NSW mission support center does not deploy forces but provides reachback capability from its operations center in Coronado, CA, to supply mission-planning support to deployed NSW units through broadband networks. The MSC collects, organizes, and disseminates mission-essential information in a form specifically tailored to the mission commander's need. Services provided include custom tactical and environmental mission-planning products, information technology (IT) help desk services, and friendly force tracking device registration and support. Aided by the MSC, mission commanders are able to focus a larger percentage of the operator's limited time on the execution phase of operations, gear preparation, and rehearsal, thereby enhancing the chance for mission success while reducing the forward footprint of NSW forces.

2.7.3 Cultural Engagement Unit

The NSW CEU provides deployable language and regional experts (LREs) and female support technicians (FSTs). It is organized into an HQ element, a training department, and six regionally aligned detachments (figure 2-8). The LREs and FSTs are embedded with deployed NSW forces to provide cultural advice and expertise and to accompany SEAL or SWCC operators to the target to assist in translation, tactical interrogation, and intelligence collection.

2.7.4 Naval Special Warfare Anchor Teams

NSW anchor teams build stability through sustained engagement within or adjacent to critical or sensitive countries. Anchor teams are organized by geographic region in which each team specializes. Anchor teams provide persistent presence and establish enduring relationships in high-priority regions by rotating the same personnel into the same locations with the same partner force.

The deploying element of the anchor team is the NSW anchor detachment—a two-man, forward-based detachment consisting of one officer and one enlisted SEAL or SWCC who conduct multiple deployments to a specific region in support of a specific host nation partner force.

2.8 NAVAL SPECIAL WARFARE GROUP ELEVEN

Commander, Naval Special Warfare Group ELEVEN is responsible for organizing, manning, training, educating, equipping, and deploying NSW Reserve Component units and personnel in support of NSW and joint special operations commanders worldwide. NSWG–11 exercises OPCON and ADCON of SEAL Teams SEVENTEEN and EIGHTEEN and advises COMNAVSPECWARCOM on NSW Reserve force matters. SEAL Teams SEVENTEEN and EIGHTEEN are a unique amalgam of both a Navy Reserve activity and a SEAL team (figure 2-9).

NWP 3-05

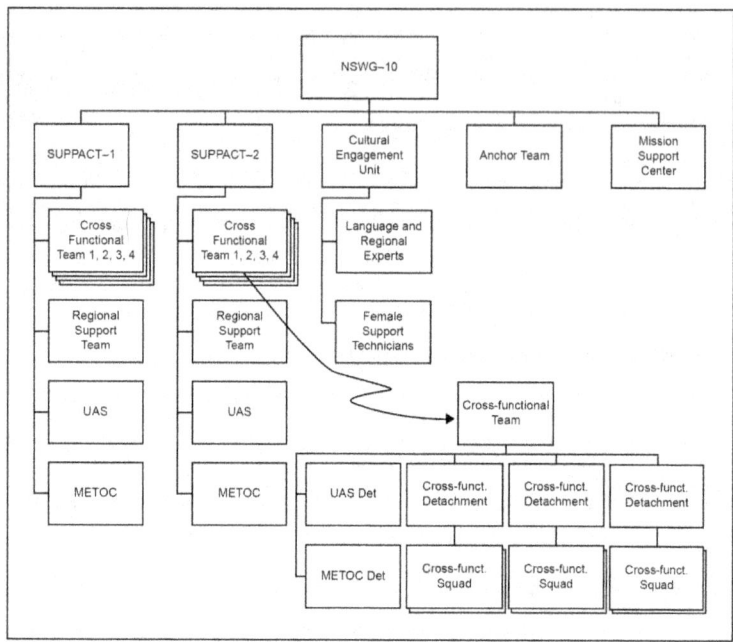

Figure 2-7. Naval Special Warfare Group TEN

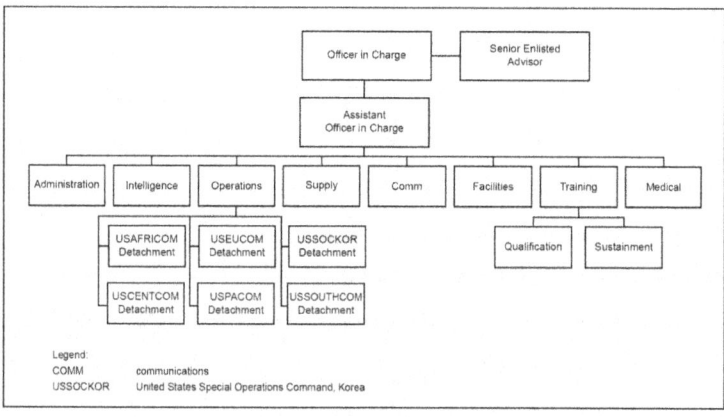

Figure 2-8. Cultural Engagement Unit

NWP 3-05

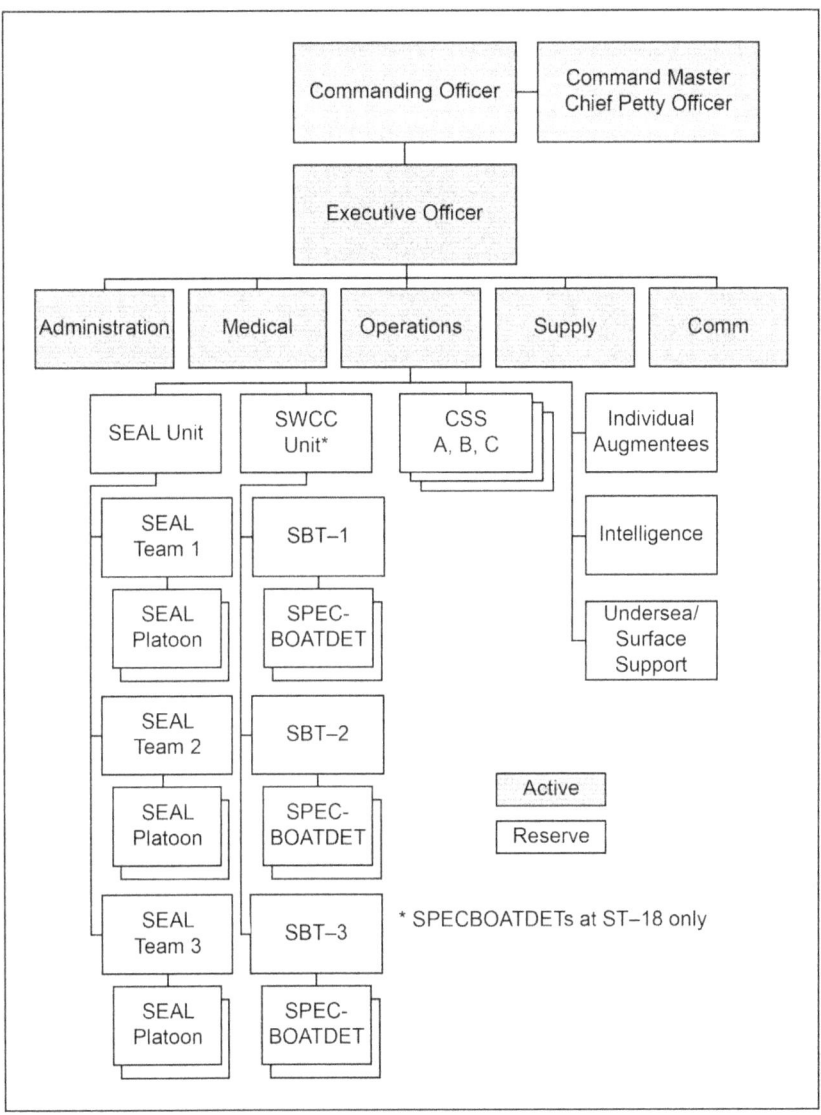

Figure 2-9. SEAL Teams SEVENTEEN and EIGHTEEN

SEAL Teams SEVENTEEN and EIGHTEEN man, train, equip, and deploy SEAL platoons, CS, and CSS in support of NSW and joint special operations commanders worldwide. SEAL Team EIGHTEEN also deploys Reserve SPECBOATDETs. They exercise ADCON of subordinate NSW Reserve units and perform administrative functions for Commander, Navy Reserve Forces Command and assigned Reserve units.

2.9 NAVAL SPECIAL WARFARE UNITS

There are five NSWUs, not to be confused with an NSWTU (naval special warfare task unit). An NSWU is a permanent forward-based NSW command and control headquarters. The NSWU has no combatant forces permanently assigned but exercises control over forces that deploy from the United States. The mission of an NSWU is to plan, coordinate, deconflict, and execute NSW operations and exercises in support of theater special operations command (TSOC) and fleet staffs. NSWU organization is discussed in chapter 4.

2.10 NAVAL SPECIAL WARFARE CENTER

Naval Special Warfare Center is the NSW proponent for training. It conducts special operations training to educate U.S. and foreign armed forces and other designated personnel in NSW tactics, techniques, procedures, and equipment. Subordinate commands are the BTC and the ATC.

2.10.1 Basic Training Command

The Basic Training Command mission is to conduct special operations training and education for U.S. and foreign armed forces and other designated personnel in basic NSW tactics, techniques, procedures, and equipment. The BTC conducts basic underwater demolition/SEAL team (BUD/S) training and SWCC basic crewman training course.

Basic Training Command also conducts SEAL qualification training for BUD/S graduates and combatant-craft crewman qualification for SWCC basic crewman training graduates. Graduates of these courses are qualified as SEAL or SWCC operators. SEAL and SWCC qualifications are discussed in chapter 4.

2.10.2 Advanced Training Command

Advanced Training Command provides standardized and accredited individual training and education for NSW operators, NSW support personnel, other USSOF, partner nation SOF, and other personnel as required, in the art and science of NSW operations. SEAL, SWCC, and support personnel attend these courses during predeployment training. Selected personnel from outside of the NSW community also attend courses when appropriate.

2.11 NAVAL SPECIAL WARFARE DEVELOPMENT GROUP

The mission of the naval special warfare development group is to provide centralized management for the test, evaluation, and development of NSW equipment and tactics, techniques, and procedures.

2.12 CENTER FOR SEAL TEAM AND SPECIAL WARFARE COMBATANT–CRAFT CREWMAN

The Center for SEAL Team and Special Warfare Combatant-Craft Crewman is a Navy echelon III command under Naval Education Training Command. CENSEALSWCC coordinates with NAVSPECWARCOM for career management, leader development, and professional military education of NSW forces.

CHAPTER 3
Command and Control

3.1 GLOBAL FORCE MANAGEMENT

Deployed forces are either assigned, attached, or allocated to a GCC who exercises COCOM, OPCON, or tactical control (TACON). When appropriate, supporting relationships, coordinating authority, and direct liaison authorized (DIRLAUTH) between forces will be specified.

The process through which forces are assigned, attached, or allocated is the Joint Chiefs of Staff (JCS) Global Force Management (GFM) process. GFM provides GCCs with tools to develop, submit, and track resourcing plans that support operational and strategic goals. GFM also supplies planners with a mechanism to input data to assignment and apportionment tables which are used in designing time-phased force and deployment data (TPFDD). Assignment and apportionment tables are part of the GFM Implementation Guidance, which the JCS release annually to provide guidance on COCOM and availability of forces.

GFM events include persistent and episodic requirements, JCS exercises, contingency plan execution, and counterterrorism (CT) execute order (EXORD) operations.

3.1.1 Persistent and Episodic Requirements

Persistent and episodic requirements are submitted by the GCCs to the joint staff as a request for forces (RFF), request for support, or as an annual baseline submission request. When there is a request for special operations forces, USSOCOM is tasked to develop a sourcing solution. In the case of requests for naval special warfare forces, USSOCOM directs NAVSPECWARCOM to conduct a feasibility assessment and recommendations for sourcing (to include an "if forced to source" solution). This analysis is presented to the Secretary of Defense (SecDef) for approval. With the exceptions of JCS exercises and CT EXORD operations, only SecDef can direct deployment of forces. If the recommended sourcing solution is approved, NSW force requirements are included in the annual SecDef-approved JCS deployment order (DEPORD) with operational control transferred to the receiving GCC. The order flows down the chain of command through a USSOCOM SOF deployment order and a NAVSPECWARCOM NAVSOF deployment order. The NAVSOF DEPORD includes the sourcing command, period of the exercise, and the number of personnel required.

The SOF DEPORD is broken into tabs, three of which are of interest to SOF: Tab B for rotational forces, Tab C for theater security cooperation plan (TSCP) events, and Tab D for joint individual requirements. Each requirement specifies, at a minimum, the requested force (e.g., naval special warfare task element (NSWTE) with one SEAL platoon), the not-to-exceed number of personnel, the earliest arrival date and latest arrival date in theater, and command relationships. Whenever forces are transferred between combatant commands, the command relationship the gaining commander will exercise (and the losing commander will relinquish) over these forces must be specified by SecDef. NSW persistent and episodic requirements include:

1. Routine force rotations. These persistent requirements include routine deployments to support the GCCs. Deployed NSW forces participate in GCC theater campaign plans and are available for crisis response. Note that, in this context, the term "crisis response" refers to an unanticipated crisis, while the term "contingency" refers to an event for which prior planning has been completed and a contingency plan exists.

2. Named operations. Named operations may be filled by forward-deployed forces or other forces provided from the United States. Examples of named operations include: ENDURING FREEDOM–Afghanistan, ENDURING FREEDOM–Philippines, and ENDURING FREEDOM–Trans Sahara.

3. Theater security cooperation plan support. These requirements can be filled by forward-deployed forces or forces provided from the United States. Examples include joint combined exchange training (JCET), counternarco-terrorism training (CNT), SFA (including NAVSCIATTS MTTs), subject matter expert exchanges, etc.

4. Joint individual augmentees. These requirements are for individuals to deploy to forward-based joint staffs to fill joint manning document billets.

3.1.2 Joint Chiefs of Staff Exercises

Requests for forces for JCS exercises fall under a separate mechanism. They are controlled by the Joint Training Information Management System and do not require SecDef approval. The chairman of the JCS is the approval authority for these exercises. Orders for NSW forces to participate in JCS exercises are received via tasking messages from COMNAVSPECWARCOM.

JCS exercise requirements may be filled by forward-deployed forces or other forces. Examples include Exercises FOAL EAGLE, ULCHI FOCUS GUARDIAN, RIMPAC, UNITAS, EMERALD WARRIOR, PANAMAX, TEMPEST WIND, JACKAL STONE, and others.

3.1.3 Contingency Plan Execution

The next category of GFM events is execution of contingency plans. The contingency plan description from JP 5-0, Joint Operation Planning, is quoted below. Note the distinction between the general term contingency plan (no acronym) and the specific term concept plan (CONPLAN).

Contingency plans are developed in anticipation of a potential crisis outside of or prior to crisis conditions. There are four levels of planning detail for contingency plans with an associated planning product for each level.

3.1.3.1 Level 1 Planning Detail—Commander's Estimate

This level of planning involves the least amount of detail and focuses on producing multiple courses of action (COAs) to address a contingency. The product for this level can be a COA briefing, command directive, commander's estimate, or a memorandum.

3.1.3.2 Level 2 Planning Detail—Base Plan

A base plan (BPLAN) describes the concept of operations, major forces, concepts of support, and anticipated timelines for completing the mission. It normally does not include annexes or TPFDD.

3.1.3.3 Level 3 Planning Detail—Concept Plan

A CONPLAN is an operation plan (OPLAN) in an abbreviated format that may require considerable expansion or alteration to convert it into an OPLAN or operation order (OPORD). It may also produce a TPFDD, if applicable.

3.1.3.4 Level 4 Planning Detail—Operation Plan

An OPLAN is a complete and detailed joint plan containing a full description of the concept of operations, all annexes applicable to the plan, and a TPFDD. It identifies the specific forces, functional support, and resources required to execute the plan and provide closure estimates for their flow into the theater.

In the case of OPLAN and CONPLAN execution, SecDef will issue an execution order that sets C-day (commence deployment) and D-day (commence operations), establishes supplemental rules of engagement, and provides direction on mobilization and TPFDD execution.

3.1.4 Counterterrorism Execute Order

Another exception to the requirement for SecDef approval is the CT EXORD. Operations in support of the CT EXORD can be directed by CDRUSSOCOM through an EXORD or the SOF DEPORD.

3.2 DEFINITIONS

3.2.1 Assigned Forces

The term "assigned" refers to those forces and resources that have been placed under the COCOM of a unified commander by the direction of SecDef in a Forces for Unified Commands memorandum. Forces are assigned or reassigned when the transfer of forces will be permanent or for an unknown period of time or when the broadest level of command authority is required or desired. All NSW forces are assigned to and under the COCOM of USSOCOM.

3.2.2 Attached Forces

The term "attached" refers to those forces and resources that have been placed under the OPCON or TACON of a GCC or other joint force commander (JFC) for a temporary situation (e.g., a 6-month deployment). This is normally accomplished through a deployment order. A DEPORD is a planning directive from SecDef, issued by the chairman of the JCS, which authorizes and directs the transfer of forces between GCCs by assignment or, more commonly, attachment.

3.2.3 Allocated Forces

Allocation is distribution of limited forces and resources for employment among competing requirements. In the GFM process, forces are allocated in the Global Force Management Allocation Plan, which is used to develop the SOF DEPORD.

3.2.4 Geographic Combatant Commander Authority

In accordance with the Forces for Unified Commands memorandum and the Unified Command Plan, except as otherwise directed by the President of the United States or SecDef, all forces operating within the geographic area assigned to a specific GCC shall be assigned or attached to and under the command of that GCC. Transient forces do not come under the chain of command of the GCC solely by their movement across operational area boundaries, except when the GCC is exercising TACON for the purpose of force protection. Unless otherwise specified by SecDef, a GCC has tactical control for exercise purposes whenever forces not assigned to that GCC undertake exercises in that GCC's area of responsibility (AOR). Note the USSOCOM exception below.

3.2.5 United States Special Operations Command Authority

To increase flexibility and agility of SOF, in 2013 changes were made to the Unified Command Plan to give USSOCOM formal responsibility to synchronize the planning, coordination, deployment, and, when directed and with the approval of the GCCs, the employment of SOF globally. COCOM of all SOF was assigned to USSOCOM through the Forces for Unified Command memorandum and assignment tables and a global SOF employment order delegated limited authority from SecDef to CDRUSSOCOM to manage and deploy special operations forces across GCC's areas of responsibility.

3.2.6 Combatant Command

COCOM is the eponymously named authority vested in the nine combatant commanders to perform all those functions of command contained in OPCON (see below) plus planning, programming, budgeting, and execution process input; assignment of subordinate commanders; relations with DOD agencies; and directive authority for logistics. COCOM cannot be delegated or transferred and is exercised exclusively by:

1. USAFRICOM
2. USCENTCOM
3. USEUCOM
4. USNORTHCOM
5. USPACOM
6. USSOUTHCOM
7. USSOCCOM
8. United States Strategic Command
9. United States Transportation Command.

3.2.7 Operational Control

OPCON is the command authority that may be exercised by commanders at any echelon and may be delegated within the command. OPCON is inherent in COCOM and is the authority to perform those functions of command over subordinate forces involving organizing and employing commands and forces, assigning tasks, designating objectives, and giving authoritative direction necessary to accomplish the mission. OPCON includes authoritative direction over all aspects of military operations and joint training necessary to accomplish missions assigned to the command.

Operational control of U.S.-based NSW forces is exercised by COMNAVSPECWARCOM. Operational control of deployed NSW forces is exercised by the GCC, normally through a TSOC, which typically delegates operational control to an naval special warfare unit or to another special operations headquarters in that theater.

3.2.8 Tactical Control

TACON is the command authority over assigned or attached forces or commands, or military capability or forces made available for tasking, that is limited to the detailed direction and control of movements or maneuvers within the operational area necessary to accomplish assigned missions or tasks. TACON is inherent in OPCON and may be delegated to and exercised by commanders at any echelon.

3.2.9 Administrative Control

ADCON is the direction or exercise of authority over subordinate or other organizations in respect to administration and support, including organization of Service forces, control of resources and equipment, personnel management, unit logistics, individual and unit training, readiness, mobilization, demobilization, discipline, and other matters not included in operational missions. ADCON may be delegated to and exercised by commanders of forces assigned to a combatant commander at any echelon at or below the level of Service component. ADCON of all NSW forces is exercised by COMNAVSPECWARCOM.

NWP 3-05

3.2.10 Support Relationships

A support relationship is established by a superior commander between subordinate commanders when one organization should aid, protect, complement, or sustain another force. This includes the Secretary of Defense designating a support relationship between GCCs as well as within a combatant command. The designation of supporting relationships is important as it conveys priorities to commanders and staffs that are planning or executing joint operations. The support command relationship is, by design, a somewhat vague but very flexible arrangement. The establishing authority (the common superior commander) is responsible for ensuring that both the supported commander and supporting commanders understand the degree of authority that the supported commander is granted. There are four defined categories of support that a GCC may direct over assigned or attached forces to ensure the appropriate level of support is provided to accomplish mission objectives. These are: general support, mutual support, direct support, and close support (see glossary for definitions).

3.2.11 Coordinating Authority

Coordinating authority is the authority delegated to a commander or individual for coordinating specific functions and activities involving forces of two or more military departments or two or more forces of the same Service. The commander or individual has the authority to require consultation between the agencies involved but does not have the authority to compel agreement. The common task to be coordinated will be specified in the establishing directive without disturbing the usual organizational relationships in other matters.

3.2.12 Direct Liaison Authorized

Direct liaison authorized is authority granted by a commander (any level) to a subordinate to directly consult or coordinate an action with a command or agency within or outside of the granting command. DIRLAUTH is more applicable to planning than operations and always carries with it the requirement of keeping the commander granting DIRLAUTH informed. DIRLAUTH is a coordination relationship, not an authority through which command may be exercised.

3.3 NAVAL SPECIAL WARFARE DEPLOYMENT ROTATION

For routine deployments, NSW forces use a 2-year, four-phase training and deployment cycle during which an 18-month interdeployment training cycle is followed by a 6-month deployment. The IDTC consists of three phases: professional development, unit-level training, and interoperability training. On successful completion of IDTC, forces are certified as ready to deploy and commence their deployment. After completion of a 6-month deployment, the cycle recommences. Details on the IDTC are available in COMNAVSPECWARCOM Instruction 3500.1, Naval Special Warfare Force Training Manual.

Contingency response and JCS exercise requirements that cannot be met by deployed forces are assigned to forces that are in IDTC, beginning with those forces in the final phase of training and working back.

3.4 NAVAL SPECIAL WARFARE TASK ORGANIZATION CHAIN OF COMMAND

Deploying NSW force packages task organized: tailored and scaled for the particular mission, situation, and conditions so that the size, seniority, and composition of the force is matched to the requirement. Depending on size and seniority, the task organization is commonly referred to as an NSWTF, NSWTG, NSWTU, or NSWTE. Because naming conventions are determined by the GCC, these organizations may be referred to by different titles depending on the operation (e.g., special operations task force, JSOTF, etc.).

Whatever the naming convention, the NSW task organization is generally assigned as the naval component of a special operations joint task force (SOJTF) or the special operations component of a Navy task force (figure 3-1). NSW also supports nonstandard chains of command in which NSW elements may be assigned to an embassy, other Government agencies, or other military organizations. In those cases where a joint special operations force is employed in a maritime area of operations, or if NSW represents a large proportion of the joint SOF capability, the senior NSW commander may be designated as the joint SOF commander (e.g., commander, JSOTF).

When working with naval forces, there are two common chains of command. The NSW task organization may be assigned as the supporting force under a Navy chain of command, e.g., an NSWTU of one or more SEAL platoons with supporting forces assigned to a ship in support of maritime interception operations (MIO) or an SDVTU assigned to support submarine operations.

Alternatively, ships, submarines, or aircraft may be assigned as a supporting force for an NSW operation under a special operations chain of command. Examples are a submarine supporting SDV operations or an afloat forward staging base (AFSB) (figure 3-1). An AFSB is a maritime platform, generally a Navy or civilian ship. When embarked, the NSW commander is designated as AFSB commander and is responsible for command and control of embarked forces and is considered the supported commander. The ship's captain, whether civilian or military, will serve as the supporting commander and is responsible for force protection, maintenance, and safety of navigation.

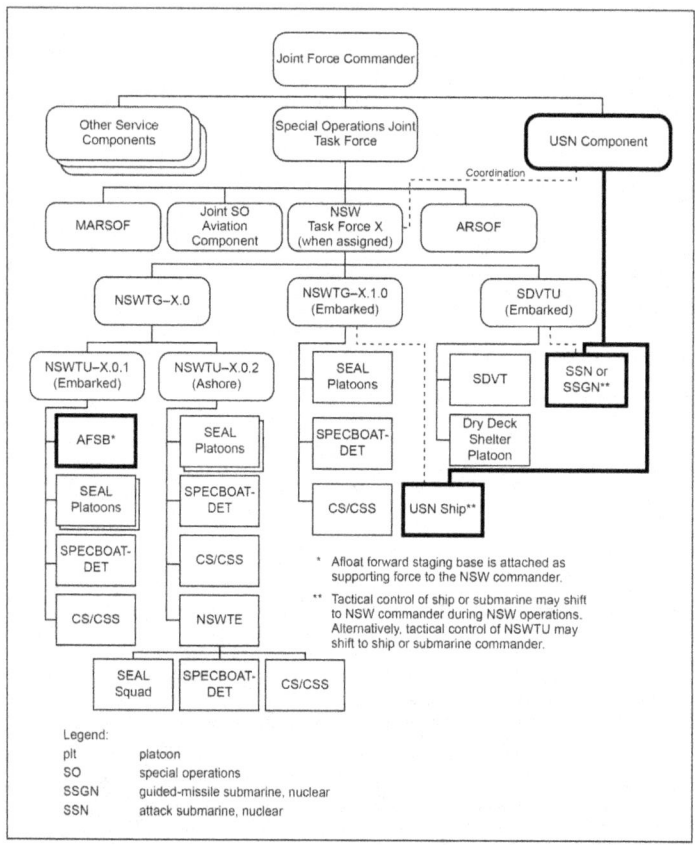

Figure 3-1. Naval Special Warfare Task Force Structure Example

3.5 THEATER SECURITY COOPERATION EVENTS

The greatest employment of NSW forces is participation in theater security cooperation events to train and assist partner nations. SEAL platoons, SPECBOATDETs, and other small specialized NSWTEs work with partner military units sharing skills and knowledge and strengthening relationships. There is not normally a requirement for a full battle staff HQ and CS and CSS requirements may be minimal. OPCON of these forces is usually assigned to the TSOC but may be delegated to other commands on a case-by-case basis.

3.6 NAVAL SPECIAL WARFARE TASK ORGANIZATION COMPONENTS

Deploying NSW force packages are built around four elements: C2, CS, CSS, and combat elements (figure 3-2). Planners must recognize that RFFs with only combat elements (e.g., SEAL platoons) result in incomplete force packages with partial capability and increased risk to mission. Most NSW operations also require Service-provided capabilities that are not organic to NSW in order to execute missions and sustain the force.

3.7 COMMAND AND CONTROL

The NSW task organization commander and headquarters are responsible for making recommendations on the proper employment of NSW forces. Details on operations of an NSW task organization HQ are available in NTTP 3-05.1, Naval Special Warfare Battle Staff Operations. Headquarters element tasks include:

1. Organizing the force. The NSW TF/TG/TU/TE commander has the authority to organize forces over which he has operational control in order to meet mission requirements. In addition to key decisions, such as basing of the HQ and subordinate units, he is also responsible for organizing the battle staff in a manner to effectively control and support subordinate NSW units. This includes coordinated communications and computer networks and administrative support and logistics support, including coordinating movement, force protection, and sustainment of forces. He may also provide liaison officers to other staffs.

2. Coordinate operational areas and control measures. The NSW commander may use a variety of means to deconflict and coordinate conventional and special operations with U.S. and Allied forces and interagency partners. There must be a well-structured process that flows between the NSW headquarters and joint and conventional force staffs and their components. This often entails use of liaison officers between staffs and emphasis on over-communication to ensure that all units have a common operational and intelligence picture. Coordination with conventional forces, establishment of procedures for support, deconfliction, and battlespace turnover require this continuing close liaison.

3. Provide operational direction and guidance. The NSW commander develops a plan for integrated employment of assigned and attached forces based on an assessment of the situation. He provides the joint commander's intent, guidance, and objectives to subordinate commanders. He provides timely and concise missions or tasks together with the role of each subordinate and supervises targeting, mission planning, integration of attached forces, and execution of operations. Using the mission command approach, tasks must be realistic and allow the subordinate as much freedom of execution as possible.

3.7.1 Naval Special Warfare Task Force

For large operations or exercises, an NSWTF HQ is established. It is commanded by a SEAL captain, often an NSWU CO or an NSWG commodore, and is sourced primarily from the NSWU or NSWG, although the exact composition varies with the requirement. The NSWTF provides command and control of two or more NSWTGs or other subordinate units.

Title	Description	Source
Headquarters Command and Control		
NSWTF C2	HQ command element	NSWG, NSWU
NSWTG C2	HQ command element	SEAL team
NSWTU C2	HQ command element	SEAL team
NSWTE C2	HQ command element	SEAL team
SDVTU C2	SDV HQ command element	SDVT–1
SPECBOAT Task Unit (SPECBOAT TU) C2	SPECBOAT TU HQ command element	SPECBOAT team
Combat Support		
Mobile communications team	Communications support	NSWG 1 and 2
DDS platoon	Submarine launch and recovery operations	SDV team 1
Cross-functional team	Intelligence gathering and analysis and UAS Det	SUPPACTs 1 and 2
Language and regional expert	Regional, cultural, and language expert	NSWG–10 CEU
FST	Female regional expert to engage partner females	NSWG–10 CEU
Combat Service Support		
Combat service support troop	Logistics, maintenance, force protection	LOGSUs 1 and 2
Undersea CSS team	Undersea logistics, maintenance, and communications	LOGSU–3
SEAL		
SEAL platoon	21-man SEAL tactical element	SEAL team
Multi-Purpose Canine Det	Dog and handler	NSWGs 1 and 2
Special Boat		
Special Boat Det-Coastal	Two coastal craft (CCL, CCM, CCH) with crew and support	SBTs 12 and 20
Special Boat Det-MCADS	Two CCM rigged for paradrop (four total craft)	SBT–20
Special Boat Det-Riverine	Two riverine craft with crews and support	SBT–22
Special Boat Det-Master Mariner	Boat crew to operate commercial and other craft	SBTs 12 and 20
SEAL Delivery Vehicle		
SDV platoon	Undersea SEAL tactical element with two SDVs	SDVT–1

Figure 3-2. Naval Special Warfare Task Organization Components

3.7.2 Naval Special Warfare Task Group

An NSWTG is commanded by a SEAL commander, generally a SEAL team CO. Although the exact composition varies with the requirement, the baseline NSWTG is a SEAL team with seven SEAL platoons, a C2 element from the same SEAL team, an MCT, a cross-functional team, a CSS team, combatant craft, tactical vehicles, rotary-wing support, NAVSOF EOD platoon, Multi-Purpose Canine, and other attachments as necessary. Six months prior to deployment, the SEAL team CO assumes operational control of the forces with which he will deploy and the task organization conducts integration training to enhance cohesion and combat effectiveness. On deployment, the SEAL team CO is designated as the NSWTG commander. The NSWTG can disaggregate its forces, typically into three or more NSWTUs at different locations to increase tactical reach. An alternative NSWTG is based around one or more SDV platoons.

Sample deployable task organizations shown in figure 3-3 give a representative number of personnel assigned to each NSW component. Note that these numbers will vary based on the requirement. For example, in a large contingency operation, an NSWTF battle staff based ashore may have 152 personnel as shown in figure 3-3, while an NSWTF based afloat may have only a dozen personnel, and a submarine-based NSWTG command and control headquarters can be as small as two people.

3.7.3 Naval Special Warfare Task Unit

An NSWTU consists of a small HQ element, one or more SEAL platoons, other operational forces as required, and CS and CSS elements for intelligence, communications, mobility, and logistics. An NSWTU is normally commanded by a SEAL task unit commander, who is a SEAL O-4. Alternatively, NSWTUs can be based around SDV platoons or boat detachments and referred to as an SDVTU or special boat task unit.

3.7.4 SEAL Delivery Vehicle Task Unit

The SDVTU, when task-organized for deployment aboard a DDS-equipped submarine, is made up of the two-man HQ element led by a SEAL lieutenant commander, one SDV platoon, one DDS platoon, and a UCSS team. An SDVTU plans and conducts undersea special operations.

3.7.5 Special Boat Task Element

When combatant craft operate independently from other NSW forces, or when there are multiple SPECBOATDETs assigned to an NSWTG, a special boat task element (SPECBOAT TE) may be designated. A SPECBOAT TE is typically subordinate to an NSWTG, although it may be assigned directly to a naval or SOF commander. It includes a small command and support element and, typically, two or more SPECBOATDETs. When necessary, the SPECBOATDET can be disaggregated into smaller maneuver elements, although—with the exception of the CCH—craft normally operate in pairs. When operating independently as an NSWTU or NSWTE, a SPECBOATDET is augmented with communications, intelligence, logistics, and mobility support.

3.7.6 Naval Special Warfare Task Element

An NSWTE is based around a small SEAL or SWCC maneuver element, generally a SEAL platoon, a SPECBOATDET, or smaller force which may be supplemented with communicators, translators, cross-functional team personnel, combat service support team personnel, or other specialists. It is commanded by a Navy lieutenant or below, warrant officer, or senior enlisted.

3.8 COMBAT SUPPORT AND COMBAT SERVICE SUPPORT

NSW provides much of its own CS and CSS support. Due to the conditions in which NSW operates, CS and CSS personnel that are assigned to NSW from fleet organizations require combat skills training prior to deployment. Most of these personnel serve in HQ support positions; other personnel accompany SEAL patrols and select personnel are trained to actively participate in combat operations.

NWP 3-05

Title	Pax	Sample NSWTU	Sample NSWTG	Sample NSWTF
Headquarters Command Group				
NSWTF C2	152	—	—	TF C2
NSWTG C2	26	—	1 TG C2	>1 TG C2
NSWTU C2	5	1 TU C2	>1 TU C2	>3 TU C2
SDVTU C2	2	As required	As required	As required
SPECBOAT-Coastal TE C2	7			
SPECBOAT-Riverine TE C2	3			
Combat Support				
MCT HQ	15	—	1 mobile comm team (42 pax: HQ plus three Det)	>1 mobile comm team
Mobile communications detachment	9	1 mobile comm Det		
DDS platoon	28	As required	As required	As required
Cross-functional team HQ	5	—	1 cross-functional team (84 pax: HQ plus three Dets plus METOC plus UAS)	>1 cross-functional team
Cross-functional Det	20	1 cross-functional Det		
METOC Det	4	As required		
UAS detachment	15	As required		
Language and regional expert	1	As required	As required	As required
FST	5	As required	As required	As required
Combat Service Support				
CSS team HQ	29	—	1 CSS team (86 pax: HQ plus three Dets)	>1 CSS team
CSS Det	19	1 CSS Det		
UCSS team	14	As required	As required	As required
SEAL				
SEAL platoon	21	2–3 SEAL platoons	4–7 SEAL platoons	>7 SEAL platoons
Multi-Purpose Canine	1	As required	As required	As required
Special Boat				
SPECBOATDET–Coastal (RIB)	14	As required	As required	As required
SPECBOATDET–Coastal (MCADS)	16			
SPECBOATDET–Riverine	17			
SPECBOATDET–Master Mariner	4			
SEAL Delivery Vehicle				
SDV platoon	16	As required	As required	As required
Navy and Other Service-provided Attachments				
Helicopter sea combat (USN) (HSC) detachment	75	As required	As required	As required
NAVSOF EOD platoon	10			
Additional attachments	TBD			
*Personnel numbers may vary				
Legend:		PAX passengers		RIB rigid inflatable boat

Figure 3-3. Sample Naval Special Warfare Task Unit/Group/Force

NWP 3-05

3.8.1 Headquarters Support

Those personnel who deploy with NSW forces but are not intended to accompany SEAL or SWCC patrols, nor participate in combat operations other than base or convoy defense, are considered HQ support. This includes personnel assigned to the NSWTF or NSWTG C2 element, CSS team personnel, and members of the MCT and cross-functional team whose primary assignment will keep them within the force protection confines of a base, ship, or submarine. Their duties may include base support, administration, logistics, maintenance, master at arms, intelligence analysis, METOC analysis, detainee operations, and communications. These personnel receive basic soldiering skills training so that they are able to participate in base or convoy defense.

3.8.2 Indirect Support

Indirect support personnel are those individuals whose skills are required at some point during a mission but who are not intended to participate in offensive combat operations. Their primary function is technical or intelligence support alongside combat elements outside the force protection confines of submarines, ships, or forward operating bases. They are integrated into SEAL or SWCC patrols and must be prepared to provide security and participate in defense. NSW indirect support personnel include select UAS personnel, language and regional experts, FSTs, and other personnel who regularly accompany combat elements as part of their primary function. Because these personnel must be integrated into a patrol formation, they receive soldiering, basic patrolling, and survival, evasion, resistance, and escape (SERE) training. During mission execution, indirect support personnel are accompanied by a SEAL or SWCC operator at all times.

3.8.3 Direct Support

Direct support are those personnel whose particular skills are necessary in the immediate proximity of the target and require them to accompany SEALs during high-risk insertions, infiltration, and final approach to the target. Direct support personnel are expected to be able to participate in offensive combat operations.

Due to the extensive training requirements necessary to accompany SEALs in offensive operations, the direct support category is limited to specialized skill sets, e.g., TIO. Direct support personnel receive additional training in physical fitness, swimming, weapons, and small-unit tactics. Training may include any facet of NSW operations, to include diving or parachute operations.

3.8.4 Undersea Combat Service Support

A variation of NSW CS and CSS are undersea combat support (UCS) and UCSS sourced from LOGSU-3. NSW UCS personnel are dive-qualified individuals whose primary mission is to provide assistance to SOF combat elements conducting dry deck shelter, lockout trunk, and lockout chamber operations. Due to the risks associated with exposure to the underwater environment, divers who conduct underway DDS operations outside the submarine hull are considered indirect support personnel.

NSW UCSS personnel are those individuals whose primary mission is to support SDV platoon training and deploy as part of an NSW task organization without actually diving or accompanying SEALs on missions. They are categorized as headquarters support personnel.

3.8.5 Service-provided Support

Navy and other Services provide additional support to the NSW task organization. This includes habitual Navy relationships with NAVSOF EOD platoons and HSC detachments and may include a wide variety of other forces.

3.9 COMBAT ELEMENTS

The combat elements of an NSW task organization are SEAL platoons, SDV platoons, SPECBOATDETs, and Multi-Purpose Canine detachments. These units are often disaggregated into smaller components (e.g., a SEAL squad) in which case, C2, CS, and CSS elements are also disaggregated into small, tailored support elements to form an NSWTE.

3.10 OTHER DEPLOYABLE NAVAL SPECIAL WARFARE FORCES

NSW deploys capabilities that are not components of an NSWTG. These forces are dedicated to train and assist operations or to preparation of the environment. This includes the NAVSCIATTS mobile training teams, NSW anchor teams, and military liaison elements.

3.11 GEOGRAPHIC COMBATANT COMMANDER ORGANIZATION

The GCC normally exercises operational control of theater SOF through a TSOC. There are currently five GCC TSOCs plus a Korean TSOC. As a JFC, the TSOC plans and conducts joint operations directed by the GCC and exercises of assigned and attached forces. For an operation or exercise, the TSOC may be designated as the joint force special operations component commander or as the SOJTF. The TSOC may establish subordinate joint task forces, such as a JSOTF, to plan and execute operations.

There is an NSWU assigned to each of the five GCC TSOCs:

1. NSWU–1, Apra Harbor, Guam: Assigned OPCON to USPACOM exercised through Special Operations Command, Pacific.

2. NSWU–2, Stuttgart, Germany: Assigned OPCON to USEUCOM exercised through Special Operations Command, Europe.

3. NSWU–3, Manama, Bahrain: Assigned OPCON to USCENTCOM exercised through Special Operations Component, United States Central Command (SOCCENT).

4. NSWU–4, Joint Expeditionary Base Little Creek–Fort Story, VA: Assigned OPCON to USSOUTHCOM exercised through Special Operations Component, United States Southern Command.

5. NSWU–10, Stuttgart, Germany: Assigned OPCON to USAFRICOM exercised through Special Operations Component, United States Africa Command.

In 2013, combatant command of NSWUs, which had formerly been exercised by the GCCs, was assigned to USSOCOM, although operational control is retained by the GCC and normally exercised through the commander of the TSOC. In most cases, the theater GCC will transfer operational control of deployed NSW forces to the NSWU commander in that theater. Administrative control of NSW forces is retained by parent NSWGs.

The following description of theater chains of command is limited to permanently established headquarters. Temporary task organizations not listed here include special operations organizations for named operations (e.g., SOF HQ for Operation ENDURING FREEDOM–Afghanistan); task organizations for specific operations (e.g., SDVTU or task group aboard submarines); execution of contingency plans which generally include establishment of a joint special operations HQ; and other provisional task organizations.

3.11.1 United States Pacific Command

NSW forces deploying to USPACOM are normally attached to NSWU-1. The CO, NSWU-1, based in Agana, Guam, has three hats.

1. CO NSWU-1.

2. Commander, Navy Special Warfare Task Force, Pacific (CNSWTF PAC). CNSWTF PAC is Commander, Special Operations Command Pacific's (CDRSOCPAC's) Navy SOF component.

3. Commander, Task Group (CTG) 78.1. CTG 78.1 is the Navy SOF component commander for Commander, SEVENTH Fleet (COMSEVENTHFLT).

Commander, United States Pacific Command (CDRUSPACOM) exercises OPCON of NSWU-1 through CDRSOCPAC. NSWU-1 is under the administrative control of CNSWG-1 and receives support from United States Pacific Fleet (figure 3-4).

3.11.2 United States European Command

NSW forces deploying to Europe are normally attached to NSWU-2. Commanding officer NSWU-2, based in Stuttgart, Germany, is designated as both Naval Special Warfare Task Group, Europe, in which capacity he is Commander, Special Operations Command, Europe's (COMSOCEUR's) naval component; and as Naval Surface Warfare Strike Group, Europe, the NSW component commander for Commander, Sixth Fleet (COMSIXTHFLT) (figure 3-5).

Commander, United States European Command (CDRUSEUCOM) exercises operational control of NSWU-2 through COMSOCEUR. NSWU-2 is under the administrative control of CNSWG-2 and receives some Service support from United States Naval Forces, Europe.

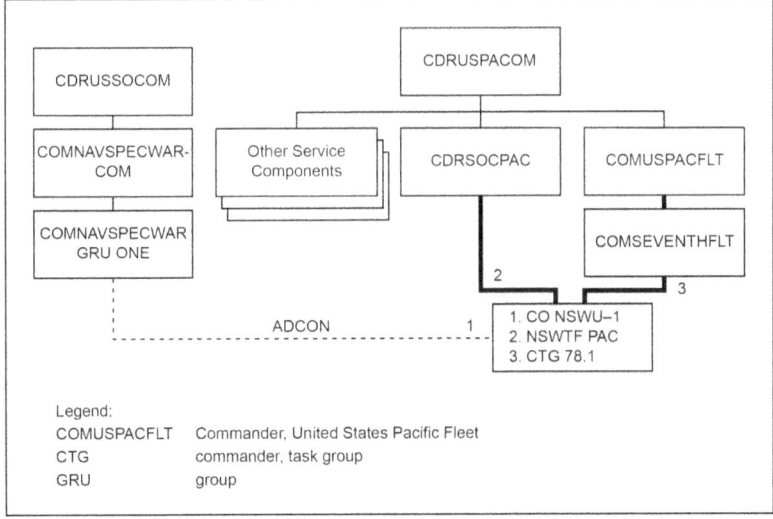

Figure 3-4. Naval Special Warfare Task Organization in Pacific Command

NWP 3-05

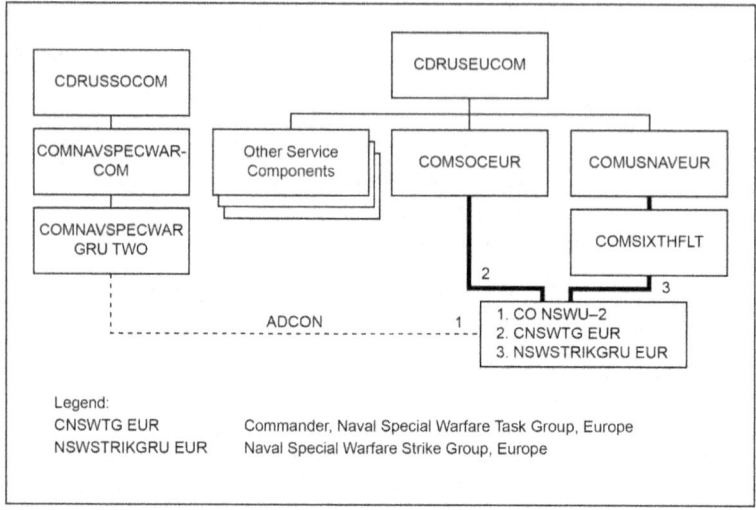

Figure 3-5. Naval Special Warfare Task Organization in European Command

3.11.3 United States Central Command

In USCENTCOM, the commanding officer of NSWU–3, based in Manama, Bahrain, serves as Commander, Joint Special Operations Task Force, Gulf Cooperation Council (JSOTF–GCC). Commander, JSOTF–GCC synchronizes all SOF TSCP engagements, partner nation SOF capability building efforts, and facilitates USSOF access and placement in the United Arab Emirates, Kuwait; Bahrain; Qatar; Oman; and the Kingdom of Saudi Arabia through persistent relationship development and key leader engagements. Commander, United States Central Command (CDRUSCENTCOM) exercises operational control of NSWU–3 through Commander, Special Operations Command, United States Central Command (COMSOCCENT) (figure 3-6). NSWU–3 is under the administrative control of CNSWG–1.

At the time of this writing, a second chain of command, the CENTCOM contingency response element (CRE), is also predominantly NSW forces. The CENTCOM CRE is a joint command based on an NSWTG and commanded by a deployed SEAL team commanding officer. The CRE's mission is to conduct special operations in support of CDRUSCENTCOM, specifically providing a rapid response force in order to prevent or counter a crisis throughout the CENTCOM AOR. Particular CRE mission sets consist of DA, SR, MIO, personnel recovery, FID, and support of contingency operations in USCENTCOM. The CRE is subordinate to SOCCENT for operational control and coordinates with the JSOTF–GCC for TSCP events.

3.11.4 United States Southern Command

Officer in charge NSWU–4, based at Joint Expeditionary Base Little Creek–Fort Story, VA, is designated as Commander, Special Operations Command, United States Southern Command's (COMSOCSOUTH's) naval component and as CTG 40.4, the NSW component commander for Commander, United States Naval Forces, Southern Command (COMUSNAVSO). Commander, United States Southern Command (CDRUSSOUTHCOM) exercises operational control of NSWU–4 through COMSOCSOUTH. NSWU–4 is under the administrative control of CNSWG–2 (figure 3-7).

NWP 3-05

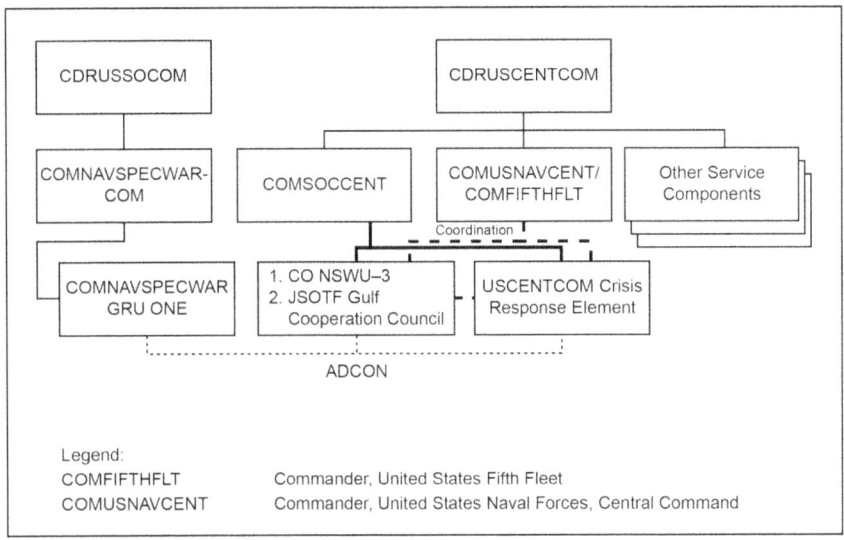

Figure 3-6. Naval Special Warfare Task Organization in Central Command

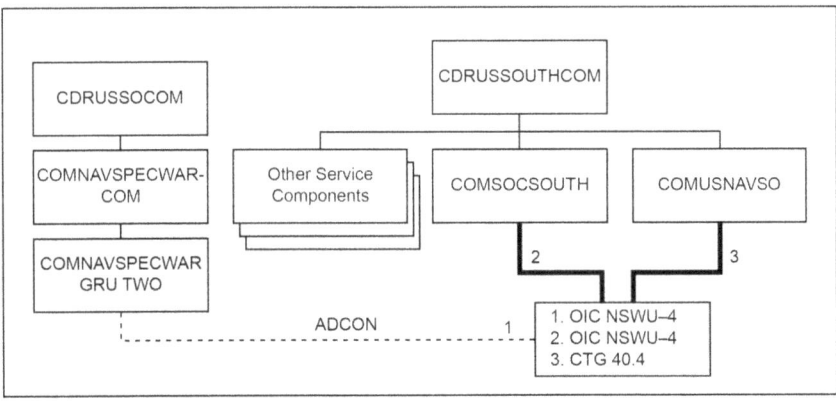

Figure 3-7. Naval Special Warfare Task Organization in Southern Command

3-15　　　MAY 2013

3.11.5 United States Africa Command

Commanding officer, NSWU–10 based in Stuttgart, Germany, is Commander, Special Operations Command, Africa's (COMSOCAFRICA) Navy SOF component. USAFRICOM exercises operational control of NSWU–10 through COMSOCAFRICA. NSWU–10 is under the administrative control of NSWG–2 (figure 3-8).

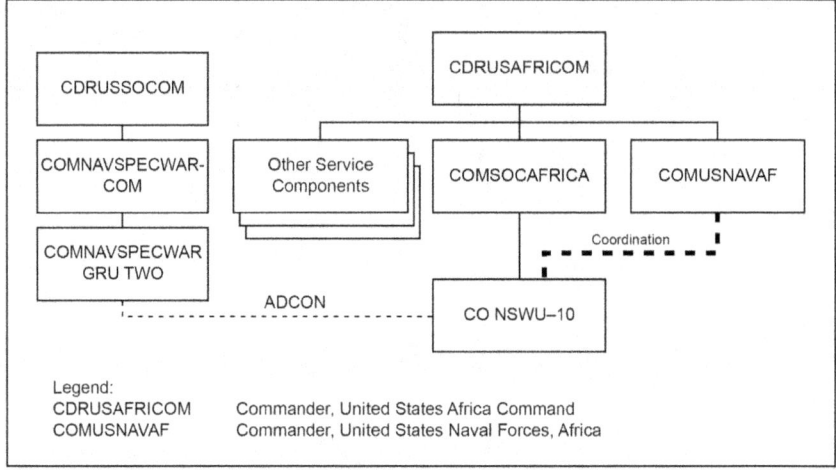

Figure 3-8. Naval Special Warfare Task Organization in Africa Command

CHAPTER 4

Naval Special Warfare Characteristics and Employment

4.1 STRATEGIC CONTEXT FOR EMPLOYMENT

While state-on-state warfare is hugely important when it occurs, it is infrequent. Meanwhile, U.S. interests are increasingly challenged by modern threats. In addition to historic competition for influence and resources (especially water and oil), there is rising volatility in the Middle East, spread of violent extremism, nuclear proliferation, threat of mass-casualty terrorism, diffusion of advanced military technologies, and increasing sophistication of transnational criminal organizations. These diverse threats may:

1. Be networked; adept at social media and cyber warfare

2. Originate from nonstate or transstate actors and can be widely—even globally—dispersed

3. Operate in or from weak or failing states exploiting under-governed spaces

4. Seek to win through psychological and political exhaustion rather than military force

5. Favor sensational violent attacks against civilian targets to foment chaos

6. Seek concealment and generate resources by blending into and operating within the population.

Strikes against these adversaries may require substantial intelligence effort to locate elusive targets, precision to limit collateral damage, stealth to achieve surprise and minimize political impact, and high confidence of success due to high political and military costs of mission failure.

The NSW force—a versatile, DA-oriented, maritime-focused organization that emphasizes innovation and flexibility, integration of intelligence, a versatile portfolio of insertion and extraction techniques, rapid response, attention to detail, and a strike capability featuring stealth, precision, and aggressive action—is well suited to these targets.

Success of military campaigns often depends less on traditional military operations and more on influencing the population and, for many modern threats, unilateral military strikes are an inappropriate or inadequate response. The future will see an emphasis on strategic partnerships to improve stability, lessen the likelihood of conflict, and prepare the United States to respond more effectively should conflict emerge. The United States will need to rely on small, politically acceptable, and regionally oriented forces operating with partner nations to proactively address challenges and mitigate threats. Strategic opportunities can be exploited in several ways.

1. Shaping. Continued presence and recurring small-scale visits to the same locations in countries with shared security interests foster enduring relationships by building trust, confidence, and understanding of critical populations. Relationships established and maintained at multiple echelons of command improve communication and information sharing, improve acceptance of U.S. presence and activities, project power, empower our allies, and strengthen security and rule of law.

2. Collaboration with partner forces. In addition to economy of force considerations, collaborative relationships enable operations that might otherwise be impossible or imprudent. This is true regardless of whether it is an independent operation conducted by a partner force with U.S. support, a combined partner-enabled operation, or a U.S. unilateral operation that is abetted by relationships with a partner country.

3. Coordination. Knowledgeable, well-connected, regionally oriented forces are able to establish relationships with theater commanders, U.S. embassy country teams, other Government agencies, and regional allies. This enhances familiarity with and access to key staff members and decision makers throughout the region in order to coordinate policies and operations.

To meet these requirements, the majority of the time and effort of deployed NSW forces is consumed by peacetime engagement and combined operations. Combined exercises, small-unit exchanges, subject matter expert exchanges, and key leader engagements contribute to a globally networked, intelligence-intensive, human-focused, regionally aware, cost-effective, and consistent influence in key locations throughout the world. NSW forces are continuously deployed and locally engaged in USCENTCOM, USPACOM, USEUCOM, USAFRICOM, and USSOUTHCOM, and are regularly employed in USNORTHCOM operations.

NSW is a relatively small force consisting of approximately 9,250 personnel: 2,700 SEALs, 700 SWCC, 750 Reserves, 4,000 CS and CSS personnel, and more than 1,100 civilians. NSW constitutes 11 percent of USSOF and less than 2 percent of Navy forces. This force costs approximately 3/10ths of a cent of every U.S. defense dollar (0.3 percent). In the context of fiscally constrained defense budgets, NSW and other USSOF have emerged as perhaps the most cost-effective U.S. weapons system.

4.2 NAVAL SPECIAL WARFARE OPERATORS

4.2.1 Navy SEAL Operator

NSW is built around a number of special skills, but the foundation of NSW operational capability and the thread that ties NSW together is the SEAL operator. The individual SEAL is a product of a selection and training process centered around basic underwater demolition/SEAL training; a rigorous, all-volunteer, 25-week course which emphasizes mental and physical toughness. Completion of BUD/S training requires proficiency in swimming, combat diving, across-the-beach operations, small-unit tactics, weapons, communications, and underwater demolitions and inculcates an enduring commitment to initiative, team work, and mission completion.

In addition to vetting of candidates prior to the commencement of training, physical and academic testing throughout the course screen out unqualified students. Moreover, there is attrition of candidates who voluntarily withdraw from the course due to arduous physical training, sleep deprivation, prolonged cold water exposure, and leadership and team work challenges. After completion of BUD/S, students move on to SEAL Qualification Training (SQT).

SQT is a 28-week sequential series of courses, including SERE; tactical air operations (static-line and freefall parachuting); helicopter rope suspension technique (rappel, fast rope, and cast and recovery); tactical combat medicine; communications; advanced special operations; language training; cold weather/mountaineering; maritime operations (small-boat operations, over-the-beach insertion); combat swimmer (closed-circuit diving, underwater ship attack); land warfare (small-unit tactics, light and heavy weapons, demolitions), combatives (military martial arts), close-quarters combat (precision close-range marksmanship, tactical decisionmaking, and ship and building clearance); and operations in a chemical, biological, radiological, and nuclear environment. The emphasis in SQT is on building and developing an individual operator capable of joining a SEAL or SDV platoon with minimal deviation in the platoon's capability.

On completion of SQT, students are awarded the Navy enlisted classification (code) of Special Warfare Operator or the officer designator of Naval Special Warfare Officer, authorized to wear the SEAL designation device (Navy SEAL trident), and assigned to a SEAL or SDV team. This selection and training process, with attrition rates exceeding 65 percent, yields SEAL operators with proven resilience, tenacity, and fortitude.

4.2.2 SEAL Platoon

On arrival at SEAL or SDV teams, SEALs are assigned to a platoon and commence 18 months of predeployment training. The 21-man SEAL platoon is the basic NSW maneuver element. Enlisted platoon personnel are assigned one or more specialties: intelligence, diving, communications, boat and vehicle operation and maintenance, ordnance, air operations, and medical. Additionally, each platoon contains specialists trained as breachers, snipers, stand-off weapons operators, advanced special operations technicians, joint terminal attack controllers (JTACs), UAS operators, tactical ground mobility operators, jumpmasters, diving supervisors, language and cultural experts, and other specialties so that each platoon is self-sufficient with all SEAL capabilities. The SEAL platoon can be disaggregated into three by seven-man SEAL squads, although each squad may not have the complete portfolio of SEAL platoon capabilities. Squads can be further divided into three- or four-man fire teams or into pairs as required.

SEAL platoon predeployment training emphasizes submerged, combatant craft, ground, and air insertion and extraction methods; patrolling and small-unit tactics; coordinated movement of maneuver elements; precision shooting and demolitions; special operations urban combat; close-quarters combat, including ship and building clearance; mission planning and rapid, decentralized, tactical decisionmaking and tactical communications; hide site selection and construction; data compression and transmission for reconnaissance missions; and sensitive site exploitation to capitalize on intelligence found on the objective and to support criminal prosecution. In addition to training in core proficiencies, SEAL platoons train in other areas of specialization to meet requirements for their deployment, e.g., advanced mountaineering and cold weather emphasis for platoons deploying to cold weather/mountainous areas. SEAL platoon operations are described in NTTP 3.05.2, SEAL Land Warfare, and NTTP 3-05.24, Naval Special Warfare-Urban Combat, Shipboarding, and Gas and Oil Platform Operations, and other references listed in References section.

4.2.3 SEAL Delivery Vehicles

A variation of the SEAL platoon is the 16-man SDV platoon in which each SEAL is a qualified SDV pilot or navigator. SDV platoons retain all of the SEAL platoon's capabilities but specialize in undersea operations. The SDV platoon can be disaggregated into two by eight-man SEAL squads, although each squad may not have the complete portfolio of SEAL platoon capabilities. SDVs are discussed in appendix A and in NTTP 3-05.42, SEAL Delivery Vehicle Tactics Handbook.

SDV operations provide more than just an alternative clandestine insertion platform. This unique capability allows planning and execution of reconnaissance and DA operations that would not be attempted by other means.

4.2.4 Special Warfare Combatant Craft–Crewman

The other specialized combatant in NSW is the SWCC, specially selected and trained personnel who operate NSW combatant craft and other craft in maritime, coastal, and riverine environments. SWCCs must complete the 7-week SWCC basic crewman training, which emphasizes physical conditioning, water competency, seamanship, navigation, boat tactics, team work, and mental toughness. The course includes a 51-hour navigation, boat tactics, and swimming evolution with little sleep and constant exposure to the elements. Completion of basic SWCC training requires proficiency in coxswain skills, over-the-horizon navigation, small-craft tactics, weapons, communications, maritime insertion and extraction, and coastal patrol and interdiction. On completion of SWCC BCT, students advance to crewman qualification training (CQT).

CQT is a 21-week course covering weapons, seamanship, first aid, navigation, communications, waterborne patrolling, marksmanship, engineering, small-unit tactics, close-quarters combat, combatives, SERE Level C, language training, and an introduction to NSW mission planning. Graduates of CQT are designated as NSW boat operators, authorized to wear the SWCC insignia, and assigned to a special boat team.

4.2.5 Versatility

In addition to screening for physical and psychological stamina, the NSW selection and qualification process fosters a self-motivated, self-reliant, problem-solving culture with an inclination to search for creative and unorthodox approaches to ambiguous problems in uncertain environments. Although tactical principles do not change, the application of those principles is in constant evolution to respond to changes in weapons, equipment, conditions, and enemy tactics. To encourage tactical and intellectual versatility and to make operations less predictable, NSW is not dogmatic about tactics. Standard tactics are used as a starting point and they are added to and expanded upon in a search for methods that are safer, faster, and simpler, developing new variations of equipment and procedures. NSW strives for creative, clever, unexpected, simple, and safe plans. Creative does not mean complex; contrarily, simplicity is the most important quality of a plan. NSW mission planning is described in NTTP 3-05.1 and NTTP 3-05.2.

Because of the inclination to improvise and the flexibility of the task organization, NSW forces have the ability, agility, and willingness to quickly task-organize into scaled and tailored force packages with a high density of the correct specialized capabilities and C2 options to plug into any size or type of operation.

4.2.6 Multi-purpose Canine

SEAL platoon capability is enhanced by the Multi-Purpose Canine, a carefully selected dog who has completed a 20-week Multi-Purpose Canine training course. Each dog is assigned a handler. The handler is a SEAL operator who trains the dog prior to deployment and remains with that dog throughout a deployment. In addition to accompanying foot, boat, or vehicle patrols, Multi-Purpose Canines can be inserted with their handler by tandem parachute, helicopter fast rope, or rappel.

The Multi-Purpose Canine's primary mission is explosive detection to provide warning of improvised explosive devices (IEDs). On indication from the dog of possible explosives, the handler marks the area where the item was found and EOD is notified to handle the threat. The Multi-Purpose Canine also possesses the ability to work on or off leash while performing other tasks, including search, locate, and apprehend enemy combatants as a nonlethal force option. In the event the target is barricaded, the dog is trained to bark alert until recalled by the handler. Multi-Purpose Canines are unable to target-discriminate, so care is used and escalation of force guidance considered prior to releasing the dog against a target.

The Multi-Purpose Canine is also employed to detect enemy weapons caches.

4.3 DIRECT ACTION ORIENTATION

NSW's organization, training, equipment, and warfighting philosophy are oriented around precision strikes that emphasize speed, stealth, and aggressive actions. A SEAL assault is characterized by clandestine approach by sea, air, or land to achieve surprise; rapid, controlled, aggressive maneuvers facilitated by clear, quick communication and decentralized command to achieve and maintain the initiative; and careful target discrimination, tactical patience, fire discipline, and precise shot placement to prevent civilian casualties and limit collateral damage.

One of the characteristics of this eyes-on, hands-on target prosecution is a degree of precision that allows commanders to exercise a measured and nuanced application of military power. This facilitates a wide range of scenario-dependent branch plans, a reasoned response to unanticipated contingencies, and allows various goals (e.g., emphasis or de-emphasis of U.S. presence) to be accentuated during mission execution.

These capabilities have been employed against a wide variety of targets and in many situations: train tunnels and bridges (Korea); radio station (Grenada); underwater boat attack (Panama); boat and oil platform assaults (Persian Gulf); enemy-held island (Persian Gulf); personnel recovery (Vietnam, Grenada, Iraq, Afghanistan); enemy supply caches (Vietnam, Afghanistan); noncombatant evacuations; ship attacks; at-sea hostage recovery; and many others. One of the mission types which capitalizes on NSW capabilities and was the predominant employment of SEALs in Vietnam, Iraq, and Afghanistan, and that has been employed in several other locations, is capture or kill missions against high-value individuals. This capability is likely to remain prominent in future irregular warfare scenarios.

4.4 MARITIME FOCUS

NSW is an extreme-environment force operating under the sea, in the open ocean, through the surf, from the air, in jungles, deserts, or high mountains; in all weather, day or night. NSW operates in all terrain but is uniquely trained and equipped to operate in maritime and riverine areas. Training and equipment emphasizes maritime insertion from aircraft, ships, submarines, SDVs, and combatant craft. All SEALs and selected SWCC, as well as certain combatant craft and other NSW equipment, are water paradrop-qualified.

Maritime attack capabilities include across-the-beach operations; near-simultaneous underway nighttime ship boarding from combatant craft and helicopter fast rope; underwater attack against ships or other targets employing closed-circuit (no bubbles) self-contained underwater breathing apparatus (SCUBA); harbor attack; and gas and oil platform assault.

This capability is significant in light of the fact that maritime nations make up 80 percent of the international community, 90 percent of trade by volume travels by sea, and there is limited access to land bases in key parts of the world.

4.4.1 Seabasing

Given the freedom of navigation laws for surface ships and the stealth of submarines, sea-based NSW forces are able to operate throughout the oceans and along the littorals of the world with few restrictions—lower profile and less intrusive presence, no reliance on host nation support, no diplomatic clearance, minimal political risk, and often without detection—to enable persistent, unobtrusive, mobile SOF presence in locations where conflict is most likely and most consequential. NSW provides special operations forces with access to the global, sustained, forward presence of the Navy.

Seabasing allows the force to position and maneuver, approach inside the horizon or retreat over it, to reposition along a coastline, to move beyond the reach of sensors, or to move to a different area as situations change. The inherent mobility increases options compared to land-based assaults from a known point, improves operational security, and mitigates an adversary's ability to evade U.S. operations. Another key value that Navy or other sea-based platforms bring is persistent presence for enduring missions such as ISR and long-term "be prepared to" missions. Seabasing can offer this presence in remote littoral areas where land bases are not available for reasons of politics, geography, or lack of supporting infrastructure. Even when land basing may be an option, sea-based platforms reduce the land-based footprint, may decrease strategic airlift requirements, and improve force security.

Seabasing support can be provided by conventional naval forces (carrier strike group, expeditionary strike group, surface strike group, missile defense surface action group, or submarines) as well as Military Sealift ships and commercial vessels. Suitable naval and United States Coast Guard (USCG) ships include guided-missile cruiser, guided-missile frigate, guided-missile destroyer, littoral combat ship, amphibious ships, submarines, joint high-speed vessel, and USCG national security cutters.

There are a variety of logistical, administrative, and C2 challenges when operating from a sea base. NSW support requirements include messing and berthing, access to planning and briefing spaces, adequate SECRET Internet Protocol Router Network (SIPRNET), Nonsecure Internet Protocol Router Network, other network connections and permissions, antenna connections, frequency management, and extra bandwidth. There may also be requirements for video feeds from airborne ISR platforms, including P3 variants and UASs. Strike groups have robust command, control, communications, computers and intelligence, surveillance, and reconnaissance (C4ISR) systems; however, these systems vary from ship to ship and may not be compatible with NSW or other SOF networks.

Other support requirements may include Judge Advocate General, public affairs, logistics, and force protection support from naval or joint forces. To maintain required on-station time during mission execution, support requirements may include at-sea resupply. Non-Navy support vessels may not have organic force protection, thus requiring defensive support.

In the history of NSW seabasing, UDT platoons and later, SEAL platoons, deployed aboard amphibious ready groups to provide hydrographic reconnaissance, advanced force operations (AFOs), and special operations support to the amphibious task force prior to amphibious landings. Since that time, the USMC has taken over routine hydrographic reconnaissance and advance force missions. In the event that a clandestine hydrographic reconnaissance is required, GCCs should request NSW support.

SEAL platoons have also deployed aboard carrier strike groups during routine 6-month deployments. These deployments, during which Navy ships spent the majority of their time on transit and training, forward presence, and sea-control missions, inhibited training and employment of NSW forces. To better employ NSW, seabasing aboard Navy ships is now limited to those periods when ships are operating in support of NSW or conducting operations (e.g., maritime interception) for which NSW capabilities may be required. Embarkation of NSW forces is arranged either in port or rendezvous at sea by boat, helicopter, or paradrop.

4.4.2 Sea-based Command and Control

When conducting operations with a carrier strike group or expeditionary strike group, a SEAL liaison officer will be assigned to the strike group staff. Whenever possible, the NSWTG or TU providing command and control of NSW forces afloat should be embarked in the same ship or submarine hosting the assault force. Pre-embarkation coordination should include planning and mission preparation space requirements (to include security requirements); C4ISR requirements; ordnance requirements; boat support (launch and recovery, fuel, deck space, maintenance space); supporting forces desired; berthing; and other requirements. Details are available in NTTP 3-07.11.1, Surface Ships in Support of Naval Special Warfare Operations; NTTP 3-07.11M, Maritime Interception Operations; NWP 3-62M, Seabasing; and NTTP 3-05.4, Naval Special Warfare Submarine Operations Manual.

4.4.3 Afloat Forward Staging Base

When a surface ship is assigned as the supporting force to SOF (SOF is the supported force), it is referred to as an afloat forward staging base.

AFSBs can range from relatively small platforms with limited ability to support an NSWTU with combat rubber raiding craft (CRRC) to larger platforms that can support a JSOTF HQ with robust combatant craft and helicopter support capabilities. Capabilities provided by the AFSB, which may consist of more than one vessel, are scaled to mission-specific requirements based on mission type and duration, threat, operational security considerations, and environmental conditions.

4.5 TARGET ACCESS

Access to a wide variety of targets is based on NSW's intelligence capability, tactical mobility, and combined action with partner forces.

4.5.1 Intelligence

Intelligence capabilities are integrated into all deployed NSW forces and include SUPPACT cross-functional teams with UAS detachments, language and regional experts, FSTs, and the NSW mission support center. Additional intelligence support is provided by individual subject matter experts and analysts from the Office of Naval Intelligence, embedded analysts on NSWTF/TG/TU staffs, and SEAL and SDV platoon intelligence personnel. Because these intelligence collection and analysis personnel are integrated into the NSW C2 organization, participate in operations working alongside SEAL and SWCC operators, and answer to the same commander, intelligence efforts are highly focused and responsive to operational needs.

There are many technical specialties included in cross-functional teams. The combined qualifications of the 84 personnel assigned to a typical team include dozens of intelligence, analysis, and technical courses and certifications in addition to basic Navy Information Dominance Corps qualifications.

The organic NSW airborne ISR capability is based on runway-independent, small-, short-, and medium-endurance UAS, including both manpack and rail-launched/cable-recovered systems. The UAS are operated primarily by UAS detachments from cross-functional teams but also by SEAL platoons and SPECBOATDETs. UAS payloads can include a variety of signals, communications, visual and infrared sensors, as well as weapons. These sensors provide real-time information, data, and full-motion video to assist in targeting, mission planning, force protection, and tactical decisionmaking during operations. Organic UAS, along with platforms provided by other forces, are employed extensively during the targeting and mission-planning process, in some cases spending hundreds of hours monitoring a single target. P–3C, MQ–1, MQ–9, and Global Hawk are particularly valuable platforms that augment organic NSW ISR. During mission execution, airborne ISR is employed tactically to monitor enemy activity and improve situational awareness.

The cross-functional team and other specially trained forces within the NSW organization conduct human intelligence, TIO (tactical electronic warfare and tactical linguists), technical surveillance, UAS operations, and gather geospatial, meteorological, oceanographic, and other information. From this information, the intelligence force conducts all-source analysis to produce tailored intelligence for NSW targeting and mission planning. In addition to traditional intelligence collection efforts emphasizing tactical information, NSW forces—particularly anchor teams and LREs—emphasize the human domain, understanding the social and cultural conditions that influence the population's attitudes and behavior.

The benefits of close integration of intelligence and operations is highlighted when targeting HVIs. The methodology to find and engage an HVI is described as an F3EAD targeting cycle. Once an HVI has been identified, multisource intelligence is used to find his general location. Over days or weeks, these data are used to build a picture of the HVI's pattern of life. With sufficient data, it may be possible to predict where the HVI will be hours or days in advance. A key part of "fixing" the HVI is finding him in a location where the risk of collateral damage is low. If this can be done, a finish operation can be conducted. The "exploit" in F3EA is the gathering and exploiting of data available immediately after the finish operation. This is referred to as sensitive site exploitation and includes biometrics, tactical questioning, and exploitation of electronic equipment. Analysis of the data and forensic investigations may lead to identification of another HVI, which begins the cycle over again. Dissemination is ensuring that the intelligence gained is properly fed back into the intelligence network.

Operations in Iraq and Afghanistan since 9/11 provided a catalyst for refinement of this targeting technique. At the beginning of the Iraq War, targeting yielded sporadic and inconsistent results. During years of missions in Iraq and Afghanistan, intelligence synthesis improved so that a jackpot became routine. In Afghanistan from May 2010 thru April 2011, special operations forces captured or killed the intended target and/or his associates 1,862 out of 2,245 times, an 83 percent success rate. The precision of this system has allowed NSW and other SOF to be highly discriminate in operations and is critical against those fleeting targets that are only rarely and briefly vulnerable.

To enhance intelligence capabilities, the NSW mission support center provides 24-hour reachback intelligence support to deployed NSW units. Services provided through the SIPRNET and other networks include sophisticated, custom-built tactical and environmental mission-planning products; IT help desk services; and friendly force tracking device registration and support. Aided by the MSC, mission commanders are able to focus a larger percentage of the operator's limited planning time on the execution phase of operations, thereby enhancing the chance for mission success. The MSC is best viewed as an extension of deployed NSW forces and reduces the number of CSS personnel required forward.

In addition to the NSW mission support center, the Navy's Office of Naval Intelligence and numerous other organizations and agencies assist with NSW target development. Sensors which may be utilized include national intelligence sensors, maritime patrol and reconnaissance aircraft, other shore-based surveillance aircraft, and tactical data from other sources. The four major conventional Navy platforms used to collect intelligence in support of special operations are ships (including ship-based UAS), submarines, P–3 variants, and land-based long-endurance UAS.

4.5.2 Tactical Mobility

Starting from a forward base, a mission can be broken down into insertion, infiltration, actions at the objective, exfiltration, and extraction. Insertion refers to the beginning of the transit up to the point where forces disembark from an insertion platform—submarine, combatant craft, helicopter, fixed-wing aircraft, or tactical vehicle—at an insertion point. Infiltration refers to the final transit to the objective—diving, swimming, combatant craft, foot patrol, mounted patrol, or helicopter. The process is reversed for exfiltration and extraction. A hallmark of NSW capabilities is its portfolio of tactical mobility, which provides seamless integration of sea, air, and land methods of insertion, infiltration, exfiltration, and extraction. The ability to operate clandestinely over, under, on, and from the sea—as well as in the air and on land—allows access to targets that may be inaccessible to other forces and is whence the acronym SEAL originates.

The ability to attack at times and locations that are unexpected is a fundamental requirement for most successful special operations. In addition to good intelligence to locate the target, this requires the ability to surreptitiously swim, dive, fly, drive, patrol, climb, or stalk in all conditions of terrain and weather and arrive at the correct location at the appointed time. This capability relies partially on tactical mobility assets that are integral to the NSW organization—combatant craft, SDVs, DDSs, and tactical ground vehicles—but also requires support from ships, submarines, and rotary- and fixed-wing aircraft. Tactical mobility is discussed in appendix A.

4.6 OTHER SUPPORT

Navy and other Services provide additional support that is not available within the NSW organization. Some of these capabilities are based on habitual Navy relationships, including NAVSOF EOD platoons and HSC detachments that are attached to the NSW task organization. Support is mission-specific and can include afloat or ashore basing, ships, submarines, rotary-wing aircraft, ISR, fire support, information operations (IO), detainee operations, interpreters, and staff support. Additional support may be provided by contractors and other Government agencies.

4.6.1 Explosive Ordnance Disposal Platoon

As part of the Navy support to naval special warfare, NAVSOF EOD platoons routinely supplement deploying NSW forces. EOD personnel accompany SEAL and SWCC operators on patrol to counter IEDs by locating, identifying, and disposing of ordnance that impedes maneuver. NAVSOF EOD platoons also provide diving support, intelligence collection, aircraft and ordnance recovery, range clearance, underwater clearance, shore detachment augmentation, and riverine operations support. Normal manning for an NAVSOF EOD platoon supporting an NSWTG is 10 men: two officers and eight enlisted.

4.6.2 Fires

NSW relies heavily on conventional forces for fire support. In addition to land-based missiles, rockets, air, artillery, and mortars, NSW employs sea-based naval fire support, including tactical air sorties from both nuclear carriers and amphibious carriers (F/A-18, AV-8B, rotary-wing), Tomahawk land-attack missiles from surface ships and SSNs, UAS, and naval surface fire support.

NSW fire support officers, ground force commanders, and JTAC plan, integrate, and execute fire support, including airborne ISR. Fire support is executed at the tactical level by the SEAL JTAC assigned to each platoon who is able to synchronize and coordinate multiple assets. Further information is available in NTTP 3-09.2.2, Naval Special Warfare Fires Supplement to MTTP for the Joint Application of Firepower.

4.6.3 Shore Basing

In many cases, deployed NSW forces require basing support. Land basing of NSW forces varies from highly developed support infrastructure to operating in austere environments with limited to nonexistent host nation or DOD support. In accordance with JP 3-05, "The logistic support of SOF units is the responsibility of their parent Service. When a theater Service component cannot satisfy its Service SOF support requirements, the GCC will determine if another Service component can do so." In practice, deployed logistics support is coordinated and provided by a combination of joint, SOF, Service, NSW organic, contractor, and commercial service providers. Missions and theaters of operation dictate the roles each of these support providers play in supporting NSW operating forces.

For rapid response or austere circumstances, NSW forces have the ability to support themselves for the first 15 days of an operation or can be self sustaining indefinitely by relying on host nation support. Normally, "Services or executive agents should be prepared to support special operations as soon as possible but not later than 15 days after SOF are employed" (JP 4-0, Joint Logistics).

4.7 COMBINED ACTION

Although the majority of NSW training is DA oriented, the majority of effort of deployed NSW forces is combined action with partner forces. The greatest contribution that can be made to a partner force is sharing knowledge and skills. Knowledge and skills taught by SEAL and SWCC operators are based on the philosophy that, rather than sending teachers who understand warfare, we send warriors who understand teaching. This enhances credibility and aids in establishing relationships. Sustained relationships with face-to-face engagements and shared experiences promote open communication and cooperation. Increased cooperation may allow information sharing and facilitate target access. When U.S. presence on target is not prudent, these relationships can facilitate judicious intelligence sharing and inconspicuous support to allow partner forces to conduct unilateral operations that further U.S. interests.

4.7.1 Combined Action Programs

Geographic combatant commander TSCPs can include several categories of U.S. Government capacity building operations: foreign military sales, foreign military finance, international military education training, FID, SFA, National Defense Authorization Act Section 1206 programs, Counterterrorism Fellowship Program, and others. Although the Department of State provides oversight and guidance for these programs, the DOD's Defense Security Cooperation Agency (DSCA) is usually responsible for implementing foreign security forces capacity building. Responsibilities and authorities for most DOD-implemented operations are coordinated by the DSCA security assistance officers who manage the programs and are located in U.S. embassies. Because the overlapping categories can be confusing, planners should consult with TSOCs and security assistance officers.

NSW and other SOF also participate in two exceptions to Department of State responsibility: joint combined exchange training (JCET) and CNT, which fall under the purview of the DOD.

4.7.2 Joint Combined Exchange Training

The JCET program is a SOF-specific training program designed to: 1) develop and enhance U.S. FID capabilities (instructional skills, foreign language proficiency, area and cultural orientation); 2) develop core mission-essential task list (METL) capabilities; and 3) support GCC theater campaign plans by training with partner nation military or paramilitary forces. The purpose of JCETs is USSOF training for FID, not to conduct FID. Accordingly, there must be a clear link between the training provided and the USSOF units' METL and USSOF must receive at least 51 percent of the benefit of the training. The JCET program is unique in that its requirements are based on SOF METLs, shaped by the theater strategy, orchestrated by theater SOCs and the U.S. ambassador, approved by SecDef in coordination with Secretary of State, and executed by the individual unit.

TSOCs working through GCCs establish JCET requirements. The host nation unit receiving the training should have SOF-like responsibilities in their country or support the host nation SOF. JCETs are episodic events, usually 2–8 weeks in duration.

The training must occur overseas unless an exception to policy is granted, referred to as a reverse JCET. When the foreign nation is a developing nation, Commander, USSOCOM may fund that nation's incremental expenses to conduct the JCET.

4.7.3 Counternarco-Terrorism Training

Counterdrug activity is focused on training host nation counterdrug forces and law enforcement agencies to detect, monitor, and interdict the cultivation, production, and trafficking of illegal drugs to the United States.

GCCs in coordination with the American embassy staff in country establish CNT requirements. In contrast to a JCET, the partner force receives 100 percent of the benefit of a CNT. CNTs are episodic events, usually limited in duration. However, there are CNTs that are 365 days in length when funding and force availability permit.

> Direct action is important, not decisive; indirect action is decisive.
>
> —*Admiral Eric Olson*

4.7.4 Regional Specialists

Partner capacity building operations are supported by NSW forces with cultural expertise, language ability of native speakers, and regional background that provide otherwise unattainable access and placement. Regional alignment improves understanding of the human domain—the local culture and how it affects behavior—and enhances relationships that improve communication, coordination, and U.S. influence. Additionally, these forces provide permanent, well-connected points of coordination with U.S. forces, U.S. embassy country teams, and other U.S. agencies in theater.

NSW area specialists include regionally oriented NSWUs, anchor team detachments, MLEs, LREs, and FSTs. Anchor team detachments and MLEs operate semi-independently, while the LREs and FSTs deploy with and are collocated with an NSWTG, TU, or TE.

Anchor team detachments and MLEs are one- to two-man elements that maintain a presence—an anchor—in select partner countries. Anchor team detachments are attached to a specific partner unit so that solid relationships are developed and maintained. They are supplemented intermittently by other NSW forces, including SEAL, SWCC, and SUPPACT cross-functional elements. Anchor team and MLE personnel are carefully screened and trained in language, regional expertise, and culture (LREC); antiterrorism/force protection and operational security; and general area preparation.

The LREs and FSTs are embedded with deployed SEAL and SWCC operators to provide cultural advice and expertise, enhance understanding of the population, and improve interface between NSW operators and local populations. LREs are Sailors who are native-born in the region of interest and are accessed into the CEU through either the Military Accessions Vital to the National Interest program or the U.S. Navy Heritage Language program. They are educated on NSW operations and operating environments and trained to function in the field next to SEAL and SWCC operators.

Female support technicians are individual female Sailors who are selected and trained to provide access to inaccessible or under-accessed female populations in partner nations who, in many targeted cultures, are isolated from exposure to male NSW personnel. This softens the operational signature and makes U.S. presence more acceptable.

4.7.5 Small-craft Training

Military to military contact is also fostered by NSW's Naval Small Craft Instruction and Technical Training School at Stennis Space Center in Mississippi. NAVSCIATTS mobile training teams and in-resident training programs prepare partner nation forces to conduct small-craft operations in riverine and littoral environments. NAVSCIATTS is well suited to provide training to and build relationships with Allied countries whose naval forces are primarily coastal.

4.8 NAVAL SPECIAL WARFARE TASKS

NSW core tasks are delineated in USSOCOM Directive 10-1, Terms of Reference for Component Commanders: Execute DA, SR, COIN, CT, SFA, and FID as primary core tasks and execute IO and UW as secondary core tasks. In accordance with USSOCOM Directive 525-16, Preparation of the Environment (PE), NSW forces also conduct PE. Many missions involve more than one of these tasks.

4.8.1 Direct Action

Direct action is defined as, "Short-duration strikes and other small-scale offensive actions conducted as a special operation in hostile, denied, or diplomatically sensitive environments and which employ specialized military capabilities to seize, destroy, capture, exploit, recover, or damage designated targets." (JP 1-02. Source: JP 3-05)

The majority of NSW combat operations are based on DA. DA can be part of FID or UW. SR missions are based primarily on DA skills and are often in support of DA. IO and PE operations are also generally conducted in conjunction with or in support of DA missions.

Direct action consists of capture or kill missions against high-value individuals; capture, neutralize, or destroy high-value or politically sensitive targets; and recovery of persons or critical items from enemy controlled areas. This includes maritime DA operations to board and seize hostile ships and gas and oil platforms and attacks against naval order of battle, ports, harbors, and supporting infrastructure. It includes sniper operations. It may include application of demolitions with precisely targeted effects in such a way as to minimize collateral damage. Demolition operations also include underwater demolitions. It may involve terminal guidance operations for air-launched munitions, naval surface fire support, artillery, and close air support. Recovery operations may include hostage or prisoner recovery, CSAR, and NEOs. Selected NSW units also conduct counterweapons-of-mass-destruction operations.

4.8.2 Special Reconnaissance

Special reconnaissance is defined as, "Reconnaissance and surveillance actions conducted as a special operation in hostile, denied, or politically sensitive environments to collect or verify information of strategic or operational significance, employing military capabilities not normally found in conventional forces." (JP 1-02. Source: JP 3-05)

Intelligence gathering is part of all NSW missions but in the case of SR, intelligence is the primary objective and enemy contact is avoided with the intent that the presence of U.S. forces is not revealed. Technological advances have developed options that include greater employment of UAS, unmanned underwater vehicles (UUVs), sensor implants, and other technical means of information collection. In addition to these systems, NSW continues to provide the option for eyes-on-target reconnaissance enhanced by advanced optics, still and video photography, and digital transmission of data which may provide information and insight about enemy activities unavailable through unmanned sensor systems. All SEALs are trained for SR, which includes clandestine tactical mobility, patrolling, hide site selection and construction, and data compression and transmission, but SEAL sniper teams are specifically trained to gather and transmit information while remaining undetected.

Maritime SR can include coastal patrol with NSW combatant craft and hydrographic reconnaissance by SEAL swimmers or SDVs. The NSW legacy mission of hydrographic reconnaissance in advance of an amphibious landing has been taken over by the USMC.

4.8.3 Counterinsurgency

Counterinsurgency is defined as "those military, paramilitary, political, economic, psychological, and civic actions taken by a government to defeat insurgency" (JP 1-02). A key aspect of COIN is the development of host nation security forces. NSW COIN operations are based on the ability to teach combat skills, regional expertise, language skills, and the ability to work among indigenous populations. COIN can include combined operations with partner forces and may include planning, ISR, mobility, communications, logistics, and other support.

4.8.4 Counterterrorism

Counterterrorism operations include actions taken directly against terrorist networks and indirectly to influence and render global and regional environments inhospitable to terrorist networks. NSW forces conduct CT operations through clandestine or low-visibility means. NSW activities within CT include, but are not limited to: capture or kill high-value individuals, intelligence operations, attacks against terrorist networks and infrastructure, hostage rescue, recovery of sensitive material, and nonkinetic activities to counter ideologies hospitable to terrorism.

4.8.5 Preparation of the Environment

PE is a term for actions to alter or shape the operational environment to create conditions favorable to the success of military operations. PE consists of operational preparation of the environment, AFOs, and intelligence operations. Refer to NTTP 3-05.5, Naval Special Warfare-PE, for more information.

4.8.6 Security Force Assistance

Security force assistance is defined as, "The Department of Defense activities that contribute to unified action by the US Government to support the development of the capacity and capability of foreign security forces and their supporting institutions." (JP 1-02). SFA supports the professionalization and the sustainable development of the capacity and capability of foreign forces, supporting institutions of host countries, and international and regional security organizations. By law, SFA operations must directly increase the capacity and/or capability of foreign security forces and/or their supporting institutions. SFA activities assist host countries to defend against internal and transnational threats; defend against external threats; contribute to coalition operations; or organize, train, equip, and advise another country's security forces or supporting institutions.

4.8.7 Foreign Internal Defense

Foreign internal defense is defined as "Participation by civilian and military agencies of a government in any of the action programs taken by another government or other designated organization to free and protect its society from subversion, lawlessness, and insurgency" (JP 3-07.1). The focus of FID efforts is to support the host nation's program of internal defense and development. FID fosters internal solutions and crosses into the nonmilitary aspects of national development with the aim of improving security and reducing instability within a nation's borders. FID can encompass any foreign entity (military, police, government, social organizations) as long as the effort is directed to internal threats (crime, terrorism, insurgency). Comparing SFA to FID, SFA is limited to military and police-type organizations but may be directed toward any threat. FID can be with any organization but must be directed against internal threats.

4.8.8 Information Operations

NSW forces do not conduct IO as a primary core activity but coordinate with naval and other forces. IO support to NSW operations consists primarily of coordinated operational security, military deception, electronic warfare, computer network operations, and military information support operations which affect enemy perceptions of friendly forces while protecting U.S. information.

NSW IO is supported by the SOF information environment, a robust and agile collection of information infrastructure, systems, processes, and knowledge that supports information dominance and provides a common environment for command, control, support, intelligence, and business functions.

4.8.9 Unconventional Warfare

NSW forces conduct UW as a secondary core activity. Often referred to as guerilla warfare, UW enables "a resistance movement or insurgency to coerce, disrupt, or overthrow a government or occupying power by operating through or with an underground, auxiliary, or guerrilla force in a denied area." (JP 1-02). UW can be conducted as part of an overall theater campaign or as an independent, subordinate campaign. When conducted independently, the primary focus of UW is on political-military and psychological objectives. When UW operations support conventional military operations, the focus shifts to primarily military objectives; however, the political and psychological implications remain. UW includes military and paramilitary aspects of resistance movements and represents the culmination of a successful effort to organize and mobilize the civil populace against a hostile government or occupying power. From the U.S. perspective, the intent is to develop and sustain these supported resistance organizations and to synchronize their activities to further U.S. national security objectives. Special operations forces assess, train, advise, and assist indigenous resistance movements to conduct UW and, when required, accompany them into combat.

4.9 GUIDELINES FOR EMPLOYMENT

NSW operations are not limited to the aforementioned tasks; in any scenario in which U.S. interests are at stake —peacetime to limited conflict to war—NSW forces can be the preferred option.

In the case of peacetime or limited conflict, NSW train and assist operations can be employed as a bulwark against instability by increasing partner capacity and improving relationships. For combat operations, decisions on employment of NSW forces are driven by target characteristics. If one or more of the following factors is significant to an operation, consideration should be given to employment of NSW.

1. Access. For hard-to-locate, hard-to-get-to targets, NSW provides intelligence capability and a broad portfolio of clandestine insertion and infiltration techniques. In situations where collaboration furthers U.S. goals, NSW can affiliate with a partner force unit giving the United States combined action alternatives. When appropriate, NSW can provide judicious intelligence sharing and other support to inconspicuously assist partner forces to conduct independent DA operations. Intelligence and strike capabilities combined in a small, agile, versatile force allow NSW to take advantage of fleeting opportunities to engage elusive targets.

2. Maritime. NSW is organized, trained, and equipped for special operations in maritime and riverine environments.

3. Stealth. NSW provides a small U.S. signature for political purposes and small tactical signature for military purposes. Stealth preserves the element of surprise.

4. Precision. NSW provides eyes-on, hands-on target discrimination as well as accuracy and judgment in target engagement. Precise, flexible, and measured application of force provides the option to emphasize nuances in mission execution to meet the commander's preferences and to respond to changes in the political and tactical situation. Precision maneuvers and shot placement increase probability of success and allow a smaller force to engage a larger force. The low risk of failure allows consideration of missions in which the political or strategic cost of failure may be high.

4.10 LIMITATIONS

Employment of NSW should also consider whether the mission is appropriate, feasible, and supportable.

1. Conservation of forces. SEAL and SWCC operators require extensive investment in selection, training, and equipment; provide unique capabilities; and are relatively few in number. They cannot be replaced quickly, nor expanded rapidly. NSW should not normally be substituted for general-purpose forces in situations where conventional forces are suitable.

2. Sustained engagement. NSW forces can deliver a high volume of weapons fire relative to their size and are designed to strike when and where least expected, employing stealth to gain surprise or use other techniques to engage the enemy with a tactical advantage. However, the small size of NSW tactical elements and limited organic supporting arms and logistics constrain their effectiveness as a static defense force or for use in sustained engagement with enemy main force units. NSW forces are well suited to seize, but often require support to hold, an objective for an extended period. NSW forces are neither trained, nor equipped to conduct sustained combat operations involving main force enemy units.

3. Timing. SEALs are a rapid-response force and can normally respond more quickly than other forces. However, preparation and rehearsal time varies with each situation. Some operations require assembly of a significant support package (submarine, ships, aircraft, etc.), extensive planning, and rehearsal. Planners should consult with the TSOC and NSW forces in theater to form a realistic timeline for mission execution.

4. Support. Support requirements may include basing, medical support, detainee operations, IO, fire support, ship, submarine, aircraft, or other attachments. Support considerations include:

    a. Transit to target area. Timely delivery of SEALs and equipment, which could include SDVs, boats, and helicopters, to a forward base.

    b. Air support. There may be a requirement for fixed-wing or rotary-wing platforms for insertion, ISR, and fire support. NSW forces have organic ISR but often require additional support from the local commander.

    c. Fire support. Planning often includes armed airborne ISR or on-call fire support for critical portions of the mission.

    d. Medical. Medical plan must include casualty evacuation to a treatment facility, often by rotary-wing.

    e. Quick reaction force. There may be a requirement for a quick reaction force or other supporting force to respond to unanticipated contingencies.

    f. Target security. There may be a requirement for general-purpose forces to provide security or to relieve NSW forces after the target is secure.

APPENDIX A
Tactical Mobility

A.1 NAVAL SPECIAL WARFARE TACTICAL MOBILITY

NSW tactical mobility encompasses the three environments of the Navy SEAL: sea, including subsurface and surface operations; air, including fixed-wing and rotary-wing operations; and land, including tactical ground mobility vehicles and foot patrol.

A.2 SUBSURFACE

Submarine lockout of an SDV or other combat submersible from a DDS followed by insertion and closed-circuit SCUBA infiltration is the most clandestine approach available and may be the only viable option to infiltrate high-threat, politically sensitive environments. Undersea insertion supports critical aspects of the national security strategy.

Submerged launch requires a submarine modified to accommodate a DDS. Large-volume SSGNs provide the ability to deploy dual DDSs and have the berthing and support facilities to sustain a large contingent of SOF operators conducting continuous operations over an extended period of time. The SSGN has storage volume for equipment and ordnance to support multiple mission profiles and the communication and C2 ability to support a small and focused JSOTF. This provides the capability to conduct a special operations undersea campaign with multiple mission sorties over an extended period of time.

With a DDS mated to the hull, a submarine can conduct a mass swimmer lockout, which is the ability to simultaneously launch a large number of SOF and CRRC while submerged. When not employing a DDS, submarines can conduct submerged lockout of SEALs and CRRC through the lockout chamber. An SSGN lockout chamber has the ability to lock out seven personnel per cycle. The SSN (Virginia class) lockout trunk can lock out 11 personnel per cycle. Alternatively, SEALs and CRRCs can be launched by briefly surfacing the submarine.

When conducting submarine lockout/-in operations, the NSWG–3 commodore is designated officer conducting the exercise (OCE). The NSWG–3 commodore can designate in writing a SEAL officer as OCE representative to assume OCE duties.

A.2.1 Dry Deck Shelter

The DDS is a submersible hangar attached to modified submarines to launch a wide variety of items, to include but not limited to: SDVs, SOF personnel, CRRC, and UUVs (figure A-1).

The DDS consists of three modules constructed as an integral unit. The first module is a hangar in which an SDV or CRRC is stowed. The second module is a transfer trunk to allow passage between the modules and the submarine. The third module is a hyperbaric recompression chamber. The interior of the DDS provides a dry working environment for mission preparations. In a typical operation, the DDS hangar module is flooded, pressurized to the surrounding sea pressure, and the hangar door opened to allow for launch and recovery. A DDS can be transported by USAF C–5/C–17 aircraft, rail, highway, or sealift to rendezvous with a submarine in port.

Figure A-1. Dual and Single Dry Deck Shelter

A.2.2 SEAL Delivery Vehicle

The SDV MK VIII MOD 1 is a "wet" submersible designed to carry combat swimmers and their cargo in fully flooded compartments (figure A-2). The SEAL pilot, navigator, and passengers are sustained by individually worn underwater breathing apparatus. Operational scenarios for the vehicle include underwater mapping and terrain exploration, location and recovery of lost or downed objects, reconnaissance missions, and DA missions. The SDV is propelled by an all-electric propulsion subsystem powered by rechargeable silver zinc batteries. Buoyancy and pitch attitude are controlled by a ballast and trim system; control in both the horizontal and vertical planes is provided through a manual control stick to the rudder, elevator, and bow planes. A computerized Doppler navigation sonar displays speed, distance, heading, altitude, and other piloting functions. SDVs can operate from forward operating sites, some surface ships (properly configured small combatants or indigenous vessels), and DDS-equipped submarines.

Current efforts are underway to replace the aging MK VIII MOD 1 with the shallow water combat submersible and to develop a dry combat submersible capability.

Figure A-2. SEAL Delivery Vehicle Launch from Dry Deck Shelter

A.3 SURFACE

Combatant craft are integral to maritime special operations. The primary mission of NSW combatant craft is to insert, extract, and support SEALs and other SOF. They are also employed in a variety of other operations: waterborne guard post, MIO, coastal patrol and interdiction, SR, master mariner operations, SFA, and FID. Combatant craft may be used to transport limited numbers of detainees from SOF objectives to afloat or ashore temporary holding facilities and for personnel recovery and noncombatant evacuation.

The primary combatant craft, NSW RIB and SOC–R, can be transported over land by vehicle line haul or organic prime movers or by sea- or airlift. NSW combatant craft may operate from any sea base or land base that is able to refuel, rearm, launch, and recover craft. Some NSW combatant craft can be paradropped.

A.3.1 Special Boat Detachments

Within the SPECBOATDET, SWCC are assigned one or more specialties: intelligence, navigation, communications, engineering, ordnance, tactical ground mobility, medical, or military freefall and air operations. Additionally, each SPECBOATDET contains specialists trained for crew-served weapons, stand-off weapons, advanced SO, master mariner, JTAC, UAS operators, tactical ground mobility, and language and cultural experts. Each SPECBOATDET is trained and proficient with all SWCC capabilities.

A.3.2 Combatant Craft–Heavy

The Sea Lion is a 77-foot stealth craft which is presently the only craft in the combatant craft-heavy (CCH) class (figure A-3).

A.3.3 Combatant Craft–Medium

The primary combatant craft-medium (CCM) is the 11-meter NSW rigid inflatable boat, a high-speed, high-buoyancy, extreme weather craft (figure A-4). The RIB is constructed of glass-reinforced plastic with an inflatable tube gunwale made of a hypalon neoprene/nylon-reinforced fabric. The 11-meter NSW RIB can carry up to an eight-man SOF element and one CRRC along with a crew of four. The RIB can be launched from land base, AFSB, or paradropped. It can be sea-lifted or air-lifted in a C–5A, C–17, or C–130.

The RIB is the workhorse of NSW but is nearing the end of its service life. Planned replacements include a 60-foot CCM and 41-foot combatant craft-assault (figure A-5).

A CCM for riverine operations and light-weather coastal operations is the 33-foot SOC–R, which can carry up to eight SEALs with seven SWCC. The SOC–R is a quick, maneuverable craft that can traverse shallow, restricted waterways while carrying large payloads. The SOC-R is designed to withstand the forces generated by operating the boat at maximum attainable speed in Sea State 2 (1 to 3-foot significant wave height) and with judicious handling in Sea State 3 (4 to 5-foot significant wave height). Characteristics of the craft include its low noise signature (virtually undetectable at idle), extreme maneuverability, and shallow draft (figure A-6).

The primary mission of the SOC–R is clandestine insertion and extraction of SEALs into a low-to-medium threat environment in a riverine arena. Secondary mission capabilities include providing fire support, serving as a staging area and reconnaissance platform, acting as a waterborne guard post, and operating as an interdiction or search and rescue craft.

Figure A-3. Sea Lion

Figure A-4. 11-meter Rigid Inflatable Boat

Figure A-5. Combatant Craft–Assault

Figure A-6. Special Operations Craft–Riverine

The SOC-R water-jet propulsion allows for shallow water operations at full throttle, providing quick extractions if the situation warrants. The large bow and stern gunwale flats are designed for quick embarkation and debarkation. The 1/4-inch aluminum-hull bottom, coupled with zero appendage drag of the water jet, allows for operating in debris-congested waters. Narrow waterways of just over 26 feet can be operated in effectively because of the craft's ability to spin about a pivot point. The SOC-R has been certified for low-velocity airdrop, internal air transport, and maritime external air transportation system (MEATS).

A.3.4 Combatant Craft–Light

The CCL is any craft that could be operated by a single operator and carry two to four passengers. The CRRC (average 15 feet in length) is an inflatable CCL capable of carrying four or five combat-equipped SEALs or other personnel for landing and recovery (insertion/extraction) on a beach/shore over short distances. It may be deck-launched/recovered from surface vessels, parachuted from various aircraft (referred to as a rubber duck drop), or cast or lowered from H-53, H-47, and H-60 helicopters. The cast technique of strapping the CRRC to the bottom of an H-60 is referred to as a Kangaroo Duck (figure A-7). CRRCs are also locked out/-in from a submarine escape trunk or DDS for launch and recovery from a submerged submarine. In a low-threat environment, a submarine may surface to launch a CRRC.

A.3.5 Master Mariner Program

A master mariner detachment is a four-man element composed of selected, mature, SWCC trained to operate a wide variety of commercial and other civilian vessels of any size. This capability can be employed overtly or clandestinely and expands flexibility of the force to operate where military craft are not available or with a negligible U.S. signature.

A.4 AIR INSERTION

Air insertion can be the quickest method to deliver SEALs and equipment to an objective area for a time-sensitive target. Water parachute drop of personnel and combatant craft can be used as an insertion method or for rendezvous with a ship or submarine at sea. All SEALs and designated SWCC and CS personnel are qualified for static line and freefall water parachute drops.

A.4.1 Maritime Craft Aerial Delivery System

MCADS is a rapid deployment and insertion method used by NSW to deliver combat-ready SEALs with RIBs quickly and safely to a water drop zone (figure A-8). In an MCADS drop, two to four RIBs and crews parachute from C-130s (one RIB each) or C-17s (two RIBs each) with SEALs and equipment. Once on the water, the boats can rendezvous with an AFSB or other ship, transit to a port or operating base, or, for a time-sensitive situation, proceed directly to the target. If refueling support is unavailable, the RIB can be refueled from an air-dropped fuel blivet.

A.4.2 Rotary Wing

Rotary-wing support can be provided by many sources from various Services and agencies and include SOF-dedicated as well as conventional fleet and other Service helicopters. Rotary-wing platforms provide precision over-land and over-water insertion and extraction, ISR, and fire support. Rotary-wing aircraft are also used by SEAL snipers as over-watch platforms for fire support and to assist in guiding tactical movement of forces during assaults.

Rotary-wing insertion techniques include parachuting, rappelling, fast roping, cast and recovery of CRRCs and SEALs, air-land insertion, and MEATS of combatant craft and SDVs (figure A-9).

Air-land refers to an operation in which the aircraft momentarily touches down to disembark SEALs and equipment, which may include vehicles.

Figure A-7. Kangaroo Duck

Figure A-8. Maritime Craft Aerial Delivery System

Figure A-9. H-60 Fast Rope Insertion

MEATS is a rapid deployment and insertion method used by NSW to deliver combat-ready SPECBOATDETs or SDV platoons into forward positions for a tactical advantage. In MEATS operations, a combatant craft can be slung beneath a CH-53 or a CH-47 for transits of up to 100 miles to a forward insertion point. Once released from the helicopter, boats can proceed directly to the objective.

The Navy's primary NSW rotary-wing forces are SO-dedicated helicopter sea combat (HSC) squadrons that train and deploy specifically in support of SOF. HSC's HH-60H and MH-60S helicopters have cabin space for assaulters, multispectral sensors for ISR, and can be configured with a variety of automatic weapons and AGM-114 Hellfire missiles for close air support. Other Navy rotary-wing assets employed by NSW include MH-60Rs flown by Navy helicopter maritime strike (HSM) squadrons to provide ISR, over-water targeting sensors, and communications packages. Best used in an airborne mission commander role, sniper platform, or antisurface strike platform, the MH-60R is limited in lift capability due to weight and space restrictions. Mine countermeasures (HM) squadrons fly the Navy's MH-53E and can provide NSW with heavy lift capability when required.

A.5 TACTICAL GROUND MOBILITY

Historically, NSW forces have employed boats, submarines, small submersibles, parachute, air-land, or foot patrol for insertions and extractions. In the 1980s, NSW first developed a tactical vehicle capability, and the capability evolved quickly in the first decade of the 2000s to deal with the threat in Iraq and Afghanistan.

NSW forces now maintain an organic tactical vehicle capability to travel long distances over land unassisted. Tactical vehicles increase the mobility, survivability, lethality, and flexibility of NSW forces. The primary mission of NSW tactical mobility assets is to transport SEAL patrols to and from insertion and extraction points in support of direct action and special reconnaissance missions. In addition to transiting to the target, vehicles can contribute fire support, security forces, quick reaction forces, casualty evacuation, and tactical convoy capabilities.

Compared to SEAL foot patrols, vehicles have greater range, speed, and payload, which allows for more fire power and longer-duration missions. Compared to air insertion, vehicles provide greater on-station time, are not restricted to landing zones or drop zones, and are generally stealthier than a helicopter. Because they are organic to NSW, these tactical vehicles are more available and more responsive to requirements than aircraft or vehicles controlled by other forces which require extra coordination and may have competing priorities. In the inland deserts and mountains of Iraq and Afghanistan, there were many cases where NSW ground mobility platforms were the only short-notice insertion method available. From 2001 to 2010, more than 90 percent of NSW missions in Iraq and Afghanistan relied, to some extent, on NSW tactical vehicles.

NSW ground mobility platforms are based on USSOCOM's family of SO vehicles and include individual, light, medium, and heavy vehicles. Individual vehicles include the mobile over snow transport, a militarized snowmobile; motorcycles; and the light tactical all-terrain vehicle (LTATV), which has been used extensively in Afghanistan (figure A-10). Light vehicles include modified commercial vehicles and the internally transportable vehicle, which is internally transportable in CH–47 and CV–22 (figure A-11). Medium vehicles include the ground mobility vehicle (GMV); and heavy vehicles include variations of the mine-resistant ambush-protected vehicle (MRAP) (figure A-12).

Makeup of a tactical ground mobility element will vary with the conditions and the requirements. A typical Afghanistan tactical ground mobility-light element supporting two SEAL platoons was four GMV and nine LTATV accompanied by two maintenance personnel. An Afghanistan tactical ground mobility-heavy element included the vehicles from the light element plus two RG33 MRAPs, two armored utility variant MRAPs, and six MRAP all-terrain vehicles with three maintenance personnel. Further information is available in NTTP 3-05.9, Naval Special Warfare Tactical Ground Mobility.

Figure A-10. Light Tactical All-terrain Vehicle

Figure A-11. Internally Transportable Vehicle

Figure A-12. RG31 A3 Mine-resistant Ambush-protected Vehicle

INTENTIONALLY BLANK

REFERENCES

JOINT

JP 3-05, Special Operations

JP 4-0, Joint Logistics

JP 5-0, Joint Operation Planning and Execution System

Secretary of Defense Letter of 11 Feb 2013, Subj: FY 2013 Forces for Unified Commands Memorandum

USSOCOM Capability Development and Programming Guidance 2015-2019 of 31 Oct 2012

USSOCOM Directive 10-1, Terms of Reference for Component Commanders

USSOCOM Directive 350-12, Joint Special Operations Task Force

USSOCOM Directive 525-16, Preparation of the Environment

USSOCOM Pub 1, Doctrine for Special Operations

U.S. NAVY

COMNAVSPECWARCOM Instruction 3500.1, NSW Force Training Manual

Memorandum of Agreement between Department of the Navy and USSOCOM of 30 April 2010

NTTP 3-05.1, Naval Special Warfare Battle Staff Operations

NTTP 3-05.1.3, Naval Special Cold Weather/Mountain Operations

NTTP 3-05.14, Naval Special Warfare Noncombatant Evacuation Operations

NTTP 3-05.15, Naval Special Warfare Unmanned Aircraft System Operations

NTTP 3-05.2, SEAL Land Warfare

NTTP 3-05.2.1, Naval Special Warfare-Breaching Operations

NTTP 3-05.21, Naval Special Warfare Combat Swimmer Operations

NTTP 3-05.23, SEAL Demolition Manual

NTTP 3-05.24, Naval Special Warfare Urban Combat, Shipboarding, and Gas and Oil Platform Operations

NTTP 3-05.3.1, Naval Special Warfare-Sniper Operations

NTTP 3-05.4, Naval Special Warfare Submarine Operations Manual

NTTP 3-05.42, SEAL Delivery Vehicle Tactics Handbook

NTTP 3-05.5, Naval Special Warfare-PE

NTTP 3-05.7, Naval Special Warfare Foreign Internal Defense

NTTP 3-05.9, Naval Special Warfare Tactical Ground Mobility

NTTP 3-06.5, Naval Special Warfare Riverine Operations

NTTP 3-07.11M, Maritime Interception Operations

NTTP 3-07.11.1, Surface Ships in Support of Naval Special Warfare Operations

NTTP 3-09.2.2, Naval Special Warfare Fires Supplement to MTTP for the Joint Application of Firepower

NWP 3-05.41, Naval Special Warfare Logistics

NWP 3-62M, Seabasing

GLOSSARY

administrative control. Direction or exercise of authority over subordinate or other organizations in respect to administration and support. (JP 1-02. Source: JP 1)

allocation. Distribution of limited forces and resources for employment among competing requirements. (JP-02. Source: JP 5-0)

apportionment. In the general sense, distribution of forces and capabilities as the starting point for planning. See also allocation. (JP 2. Source: JP 5-0)

assigned forces. Those forces and resources that have been placed under the combatant command (command authority) of a unified commander in the Forces for Unified Commands Memorandum. Forces and resources so assigned are available for normal peacetime operations of that command. (NTRP 1-02)

clandestine operation. An operation sponsored or conducted by governmental departments or agencies in such a way as to assure secrecy or concealment. A clandestine operation differs from a covert operation in that emphasis is placed on concealment of the operation rather than on concealment of identity of the sponsor. In special operations, an activity may be both covert and clandestine and may focus equally on operational considerations and intelligence-related activities. (JP 1-02. Source: JP 3-05.1)

close support. That action of the supporting force against targets or objectives which are sufficiently near the supported force as to require detailed integration or coordination of the supporting action with the fire, movement, or other actions of the supported force. (JP 1-02. Source: JP 3-31)

combatant command (command authority). Nontransferable command authority, which cannot be delegated, of a combatant commander to perform those functions of command over assigned forces involving organizing and employing commands and forces; assigning tasks; designating objectives; and giving authoritative direction over all aspects of military operations, joint training, and logistics necessary to accomplish the missions assigned to the command. (JP 1-02. Source: JP 1)

combat service support (CSS). The essential capabilities, functions, activities, and tasks necessary to sustain all elements of operating forces in theater at all levels of war. Within the national and theater logistic systems, it includes but is not limited to that support rendered by service forces in ensuring the aspects of supply, maintenance, transportation, health services, and other services required by aviation and ground combat teams to permit those units to accomplish their missions in combat. Combat service support encompasses those activities at all levels of war that produce sustainment to all operating forces on the battlefield. (JP 1-02. Source: JP 4-0)

combat support (CS). Fire support and operational assistance provided to combat elements. (JP 1-02. Source: JP 4-0)

combined. (DOD) Between two or more forces or agencies of two or more allies. (When all allies or services are not involved, the participating nations and services shall be identified, e.g., combined navies.) (JP 1-02.)

coordinating authority. The commander or individual who has the authority to require consultation between the specific functions or activities involving forces of two or more Services, joint force components, or forces of the same Service or agencies, but does not have the authority to compel agreement. (JP 1-02. Source: JP 1).

counterinsurgency. Comprehensive civilian and military efforts taken to defeat insurgency and to address any core grievances. (JP 1-02. Source: JP 3-24)

counterterrorism. Actions taken directly against terrorist networks and indirectly to influence and render global and regional environments inhospitable to terrorist networks. (JP 1-02. Source: JP 3-26)

covert operation. An operation that is so planned and executed as to conceal the identity of or permit plausible denial by the sponsor. See also clandestine operation; overt operation. (JP 1-02. Source: JP 3-05)

cross-functional team. An NSWG-10 ISR/PE-centric element focused on the fusion of intelligence collection and analysis functions to provide organic NSW support for the find, fix, finish, exploit, analyze, and disseminate (F3EAD) targeting cycle. (NTRP 1-02)

deconflict. To prevent a hostile encounter between forces, usually between friendly forces in the same area of operations. (NTRP 1-02)

direct action. Short-duration strikes and other small-scale offensive actions conducted as a special operation in hostile, denied, or diplomatically sensitive environments and which employ specialized military capabilities to seize, destroy, capture, exploit, recover, or damage designated targets. (JP 1-02. Source: JP 3-05)

direct liaison authorized. That authority granted by a commander (any level) to a subordinate to directly consult or coordinate an action with a command or agency within or outside of the granting command. (JP 1-02. Source: JP 1)

direct support. A mission requiring a force to support another specific force and authorizing it to answer directly to the supported force's request for assistance. (JP 1-02. Source: JP 3-09.3)

exfiltrate. The movement of personnel or units from areas under enemy control by stealth, deception, surprise, or clandestine means. (NTRP 1-02)

extraction. The removal, by some type of platform, of personnel or units from areas under enemy control. (NTRP 1-02)

female support technicians. Specially vetted and trained Navy female combat support personnel.

foreign internal defense (FID). Participation by civilian and military agencies of a government in any of the action programs taken by another government or other designated organization to free and protect its society from subversion, lawlessness, insurgency, terrorism, and other threats to its security. (JP 1-02. Source: JP 3-22)

general support. That support which is given to the supported force as a whole and not to any particular subdivision thereof. (JP 1-02. Source: JP 3-09.3)

global force management. A process that provides near-term sourcing solutions while providing the integrating mechanism between force apportionment, allocation, and assignment. (JP 1-02. Source: JP 3-35)

infiltration. 1. The movement through or into an area or territory occupied by either friendly or enemy teams or organizations. The movement is made, either by small groups or by individuals, at extended or irregular intervals. When used in connection with the enemy, it implies that contact is avoided. 2. In intelligence usage, placing an agent or other person in a target area in hostile territory. Usually involves crossing a frontier or other guarded line. Methods of infiltration are: black (clandestine); gray (through legal crossing but under false documentation); white (legal). (JP 1-02. Source: JP 3-05.1)

information operations. The integrated employment during military operations, of information-related capabilities, in concert with other lines of operation, to influence, disrupt, corrupt, or usurp the decisionmaking of adversaries and potential adversaries while protecting our own. (JP 1-02. Source: JP 3-13)

insertion. 1. Placement of teams and equipment into an operational area in air assault operations. 2. Placement of observation posts, patrols, or raiding parties by helicopter, parachute, watercraft, or other means. Stealth is normally desired in the execution of an insertion. (NTRP 1-02)

insurgency. The organized use of subversion and violence by a group or movement that seeks to overthrow or force change of a governing authority. Insurgency can also refer to the group itself. (JP 1-02. Source: JP 3-24)

intelligence. The product resulting from the collection, processing, integration, evaluation, analysis, and interpretation of available information concerning foreign nations, hostile or potentially hostile forces or elements, or areas of actual or potential operations. The term is also applied to the activity which results in the product and to the organizations engaged in such activity. (JP 1-02. Source: JP 2-0)

joint. Connotes activities, operations, organizations, etc., in which elements of two or more Military Departments participate. (JP 1-02. Source: JP 1)

joint force commander. A general term applied to a combatant commander, subunified commander, or joint task force commander authorized to exercise combatant command (command authority) or operational control over a joint force. (JP 1-02. Source: JP 1)

joint special operations task force. A joint task force composed of special operations units from more than one Service, formed to carry out a specific special operation or prosecute special operations in support of a theater campaign or other operations. (JP 1-02. Source: JP 3-05)

logistics support unit. Provides administrative, maintenance, logistical, and medical support to a naval special warfare group and its subordinate sea-air-land teams. Abbreviated as LOGSU.

low visibility operations. Sensitive operations wherein the political-military restrictions inherent in covert and clandestine operations are either not necessary or not feasible; actions are taken as required to limit exposure of those involved and/or their activities. Execution of these operations is undertaken with the knowledge that the action and/or sponsorship of the operation may preclude plausible denial by the initiating power. (JP 1-02. Source: JP 3-05.1)

maritime interception operations. Efforts to monitor, query, and board merchant vessels in international waters to enforce sanctions against other nations such as those in support of United Nations Security Council Resolutions and/or prevent the transport of restricted goods. (JP 1-02. Source: JP 3-0)

military information support operations. Planned operations to convey selected information and indicators to foreign audiences to influence their emotions, motives, objective reasoning, and ultimately the behavior of foreign governments, organizations, groups, and individuals in a manner favorable to the originator's objectives. (JP 1-02. Source: JP 3-13.2)

mobile training team. A team consisting of one or more U.S. military or civilian personnel sent on temporary duty, often to a foreign nation, to give instruction. The mission of the team is to train indigenous personnel to operate, maintain, and employ weapons and support systems, or to develop a self-training capability in a particular skill. The Secretary of Defense may direct a team to train either military or civilian indigenous personnel, depending upon host-nation requests. (JP 1-02)

mutual support. That support which units render each other against an enemy, because of their assigned tasks, their position relative to each other and to the enemy, and their inherent capabilities. (JP1-02. Source: JP 3-31)

naval special warfare. A naval warfare specialty that conducts operations with emphasis on maritime, coastal, and riverine environments using small, flexible, mobile units operating under, on, and from the sea. (JP 1-02. Source: JP 3-05)

naval special warfare group (NSWG). A permanent Navy echelon III major command to which most naval special warfare forces are assigned for some operational and all administrative purposes. It consists of a group headquarters with command and control, communications, and support staff; sea-air-land teams; and sea-air-land team delivery vehicle teams (JP 1-02. Source: JP 3-05.1)

naval special warfare task element (NSWTE). A temporary, task-organized operational component of a naval special warfare task unit, tailored to a specific operational requirement (e.g., all or part of a SEAL team platoon and all or part of a special boat detachment, etc.) (NTRP 1-02)

naval special warfare task force (NSWTF). A provisional naval special warfare organization that plans, conducts, and supports special operations in support of joint special operations task force commanders and fleet commanders. It may be the naval component of a joint special operations task force, or its commander may command the joint special operations task force. It is normally commanded by a naval special warfare unit commanding officer or a naval special warfare group commander. It provides command and control of one or more naval special warfare task groups. Within a theater, this may be a standing organization, an organization created to address a unique, emergent operational requirement, or an organization activated only for planning, exercises, or contingencies. (NTRP 1-02)

naval special warfare task group (NSWTG). A provisional naval special warfare organization that plans, conducts, and supports special operations in support of joint force special operations component commanders and/or fleet commanders. It comprises a command and control element and one or more subordinate naval special warfare task units. It is often commanded by a SEAL team commander (O-5), normally the commanding officer of a deployed naval special warfare squadron. (NTRP 1-02)

naval special warfare task unit (NSWTU). A provisional naval special warfare organization and subordinate unit of a naval special warfare task group. Normally commanded by a SEAL team lieutenant commander (O-4), a naval special warfare task unit consists of a command and control element, an operational element, and other attachments, e.g., mobility element. (NTRP 1-02)

naval special warfare unit (NSWU). A permanent Navy organization under the combatant command (command authority) of a geographic combatant commander that provides for command and control and support of designated naval special warfare forces under the operational control of that geographic combatant commander. (NTRP 1-02)

noncombatant evacuation operations. Operations directed by the Department of State or other appropriate authority, in conjunction with the Department of Defense, whereby noncombatants are evacuated from foreign countries when their lives are endangered by war, civil unrest, or natural disaster to safe havens as designated by the Department of State. See also evacuation; geospatial-intelligence contingency package; noncombatant evacuees; operation; safe haven. (JP 1-02. Source: JP 3-68)

operational control (OPCON). The authority to perform those functions of command over subordinate forces involving organizing and employing commands and forces, assigning tasks, designating objectives, and giving authoritative direction necessary to accomplish the mission. (JP 1-02. Source: JP 1)

preparation of the environment. An umbrella term for operations and activities conducted by selectively trained special operations forces to develop an environment for potential future special operations. Also called PE. (JP 1-02. Source: JP 3-05)

reconnaissance. A mission undertaken to obtain, by visual observation or other detection methods, information about the activities and resources of an enemy or adversary or to secure data concerning the meteorological, hydrographic, or geographic characteristics of a particular area. (JP 1-02. Source: JP 2-0)

SEAL delivery vehicle platoon (SDV PLT). The deployable, operational entity of a SEAL delivery vehicle team. A SEAL delivery vehicle platoon consists of a SEAL delivery vehicle, an operational (SEAL) element, and support personnel. (NTRP 1-02)

SEAL delivery vehicle team (SDVT). US Navy special operations command, normally commanded by a SEAL team commander (O-5), that organizes, trains, and equips Navy special operations forces to operate SEAL delivery vehicle platoons, dry deck shelter platoons, and advanced SEAL delivery system task units. (SEAL delivery vehicle team ONE only). (NTRP 1-02)

SEAL team. US Navy special operations unit or command, normally commanded by a SEAL team commander (O-5), that organizes, trains, and equips Navy special operations force to conduct special operations with a focus on operations in maritime, littoral, and riverine environments. (NTRP 1-02)

security assistance. Group of programs authorized by the Foreign Assistance Act of 1961, as amended, and the Arms Export Control Act of 1976, as amended, or other related statutes by which the United States provides defense articles, military training, and other defense-related services, by grant, loan, credit, or cash sales in furtherance of national policies and objectives. Security Assistance is an element of Security cooperation funded and authorized by Department of State to be administered by Department of Defense/Defense Security Cooperation Agency. (JP 1-02. Source: JP 3-22)

security cooperation. All Department of Defense interactions with foreign defense establishments to build defense relationships that promote specific US security interests, develop allied and friendly military capabilities for self-defense and multinational operations, and provide US forces with peacetime and contingency access to a host nation. (JP 1-2. Source: JP 3-22)

security force assistance. The Department of Defense activities that contribute to unified action by the US Government to support the development of the capacity and capability of foreign security forces and their supporting institutions. Also called SFA. (JP 1-02. Source: JP 3-22)

special boat team (SBT). US Navy special operations command, normally commanded by a SEAL team commander (O-5), that organizes, trains, and equips Navy special operations forces to conduct or support special operations employing combatant and other craft. (NTRP 1-02)

special boat team (SBTP). The deployable, operational entity of a special boat team normally commanded by a SEAL team lieutenant (O-3) and consisting of a small command and control element, combat service support and other capabilities, and one or more special boat detachments. (NTRP 1-02)

special operations. Operations requiring unique modes of employment, tactical techniques, equipment, and training often conducted in hostile, denied, or politically sensitive environments and characterized by one or more of the following: time sensitive, clandestine, low visibility, conducted with and/or through indigenous forces, requiring regional expertise, and/or a high degree of risk (JP 1-02. Source: JP 3-05)

special operations forces. Those active and Reserve Component forces of the Military Services designated by the Secretary of Defense and specifically organized, trained, and equipped to conduct and support special operations. (JP 1-02. Source: JP 3-05.1)

special reconnaissance. Reconnaissance and surveillance actions conducted as a special operation in hostile, denied, or politically sensitive environments to collect or verify information of strategic or operational significance, employing military capabilities not normally found in conventional forces. (JP 1-02. Source: JP 3-05)

special warfare combatant-craft crewman (SWCC). US Navy special operations personnel that operate combatant and other craft to conduct special operations. (NTRP 1-02)

support. An element of a command that assists, protects, or supplies other forces in combat. (JP 1-02. Source: JP 1)

tactical control. Command authority over forces that is limited to the detailed direction and control of movements or maneuvers within the operational area necessary to accomplish missions or tasks assigned. (JP 1-02. Source: JP 1)

unconventional warfare. Activities conducted to enable a resistance movement or insurgency to coerce, disrupt, or overthrow a government or occupying power by operating through or with an underground, auxiliary, and guerrilla force in a denied area. (JP 1-02. Source: JP 3-05)

LIST OF ACRONYMS AND ABBREVIATIONS

ADCON	administrative control
AFO	advanced force operations
AFSB	afloat forward staging base
AOR	area of responsibility
ARSOF	Army special operations forces
ATC	Advance Training Command
BPLAN	base plan
BSU	boat support unit
BTC	Basic Training Command
BUD/S	basic underwater demolition/SEAL team
C2	command and control
C4ISR	command, control, communications, computers and intelligence, surveillance, and reconnaissance
CCDR	combatant commander
CDRSOCPAC	Commander, Special Operations Command Pacific
CDRUSAFRICOM	Commander, United States Africa Command
CDRUSCENTCOM	Commander, United States Central Command
CDRUSEUCOM	Commander, United States European Command
CDRUSPACOM	Commander, United States Pacific Command
CDRUSSOCOM	Commander, United States Special Operations Command
CDRUSSOUTHCOM	Commander, United States Southern Command
CENSEALSWCC	Center for SEAL Team and Special Warfare Combatant Craft-Crewman
CEU	cultural engagement unit
CNO	Chief of Naval Operations
CNT	counternarco-terrorist training

NWP 3-05

CNSWG	Commander, Naval Special Warfare Group
CNSWTF	commander, naval special warfare task force
CNSWTF–PAC	Commander, Naval Special Warfare Task Force, Pacific
CNSWTG	commander, naval special warfare task group
CO	commanding officer
COA	course of action
COCOM	combatant command (command authority)
COIN	counterinsurgency
COMM	communications
COMNAVCENT	Commander, United States Naval Forces, Central Command
COMNAVSPECWARCOM	Commander, Naval Special Warfare Command
COMSEVENTHFLT	Commander, Seventh Fleet
COMSIXTHFLT	Commander, Sixth Fleet
COMSOCAFRICA	Commander, Special Operations Command, Africa
COMSOCCENT	Commander, Special Operations Command, United States Central Command
COMSOCEUR	Commander, Special Operations Command, Europe
COMSOCSOUTH	Commander, Special Operations Command, United States Southern Command
COMUDU	commander, underwater demolition unit
COMUSNAVSO	Commander, United States Naval Forces, Southern Command
COMUSPACFLT	Commander, United States Pacific Fleet
CONPLAN	concept plan
CP&I	coastal patrol and interdiction
CQT	crewman qualification training
CRRC	combat rubber raiding craft
CRE	contingency response element
CS	combat support
CSAR	combat search and rescue

CSS	combat service support
CT	counterterrorism
CTG	commander, task group
DA	direct action
DDS	dry deck shelter
DEPORD	deployment order
Det	detachment
DEVGRU	development group
DEVRON	development squadron
DIRLAUTH	direct liaison authorized
DOD	Department of Defense
DSCA	Defense Security Cooperation Agency
EOD	explosive ordnance disposal
F3EA	find, fix, finish, exploit, and analyze
F3EAD	find, fix, finish, exploit, analyze, and disseminate
FID	foreign internal defense
FST	female support technician
GCC	geographic combatant commander
GFM	Global Force Management
GMV	ground mobility vehicle
GRU	group
HM	helicopter mine countermeasures squadron
HQ	headquarters
HSC	helicopter sea combat (USN)
HSM	helicopter maritime strike squadron
HVI	high-value individual
IDTC	interdeployment training cycle
IED	improvised explosive device

IO	information operations
ISR	intelligence, surveillance, and reconnaissance
IT	information technology
JCET	joint combined exchange training
JCF	Joint Chiefs of Staff
JFC	joint force commander
JSOTF	joint special operations task force
JSOTF–GCC	Joint Special Operations Task Force, Gulf Cooperation Council
JTAC	joint terminal attack controller
JTF	joint task force
LOGSU	logistics support unit
LRE	language and regional expert
LTATV	light tactical all-terrain vehicle
LTJG	lieutenant junior grade
MARSOF	Marine Corps special operations forces
MCAD	maritime craft aerial delivery system
MEATS	maritime external air transportation system
METL	mission-essential task list
METOC	meteorological and oceanographic
MIO	maritime interception operations
MISO	military information support operations
MSC	mission support center
MST	mobile support team
MTT	mobile training team
NAVSCIATTS	Naval Small Craft Instruction and Technical Training School
NAVSOF	Navy special operations forces
NAVSPECWARCOM	Naval Special Warfare Command
NCDU	naval combat demolition unit

NEO	noncombatant evacuation operation
NSW	naval special warfare
NSWC	Naval Special Warfare Center
NSWG	naval special warfare group
NSWTE	naval special warfare task element
NSWTF	naval special warfare task force
NSWTG	naval special warfare task group
NSWTG EUR	Naval Special Warfare Task Group, Europe
NSWTU	naval special warfare task unit
NSWU	naval special warfare unit
OCE	officer conducting the exercise
OIC	officer in charge
OPCON	operational control
OPLAN	operation plan
PB	patrol boat
PE	preparation of the environment
PLT	platoon
PTF	patrol torpedo fast (naval attack craft)
RFF	request for forces
RIB	rigid inflatable boat
S&R	scouts and raiders
SBU	special boat unit
SCUBA	self-contained underwater breathing apparatus
SDV	SEAL delivery vehicle
SDVTU	SEAL delivery vehicle task unit
SEAL	sea-air-land
SecDef	Secretary of Defense
SERE	survival, evasion, resistance, and escape

SFA	security force assistance
SIPRNET	SECRET Internet Protocol Router Network
SOCCENT	Special Operations Component, United States Central Command
SOC-R	special operations craft-riverine
SOF	special operations forces
SPECBOAT	special boat
SPECBOATDET	special boat detachment
SPECBOAT TE	special boat task element
SPECBOAT TU	special boat task unit
SQT	SEAL qualification training
SR	special reconnaissance
SSGN	nuclear-powered guided missile submarine
SSN	nuclear-powered attack submarine
SUPPACT	support activity
SWCC	special warfare combatant-craft crewman
TAC D&E	tactical development and evaluation
TACON	tactical control
TF	task force
TIO	tactical information operations
TPFDD	time-phased force and deployment data
TRADET	training detachment
TSCP	theater security coordination plan
TSOC	theater special operations command
TTP	tactics, techniques, and procedures
UAS	unmanned aircraft system
UCS	undersea combat support
UCSS	undersea combat service support
UDT	underwater demolition team

USAF	United States Air Force
USAFRICOM	United States Africa Command
USCENTCOM	United States Central Command
USEUCOM	United States European Command
USMC	United States Marine Corps
USN	United States Navy
USNORTHCOM	United States Northern Command
USPACOM	United States Pacific Command
USSOCOM	United States Special Operations Command
USSOF	United States special operations forces
USSOUTHCOM	United States Southern Command
UUV	unmanned underwater vehicle
UW	unconventional warfare
WWII	World War II

INTENTIONALLY BLANK

NWP 3-05

LIST OF EFFECTIVE PAGES

Effective Pages	Page Numbers
MAY 2013	1 thru 16
MAY 2013	EX-1, EX-2
MAY 2013	1-1 thru 1-8
MAY 2013	2-1 thru 2-12
MAY 2013	3-1 thru 3-16
MAY 2013	4-1 thru 4-14
MAY 2013	A-1 thru A-12
MAY 2013	Reference-1, Reference-2
MAY 2013	Glossary-1 thru Glossary-6
MAY 2013	LOAA-1 thru LOAA-8
MAY 2013	LEP-1, LEP-2

INTENTIONALLY BLANK

NWP 3-05
MAY 2013

www.ingramcontent.com/pod-product-compliance
Lightning Source LLC
Chambersburg PA
CBHW050117230526
45470CB00004B/1877